DEDICATION

To
Bonnie Lou
and our Children:
Michael, Evan, Stephen,
Jeffrey, Tracy and Rebecca

TABLE OF CONTENTS

Dedication . . . .*iii*
Introduction . . . .*viii*
Abbreviations . . . .*xii*

PART ONE: GOD AND HIS CREATIONS
CHAPTER ONE / THE UNIVERSES 3
    The Immensity of Creation 3
    Three Cosmogonies 6
    The Eternal Universe 8

CHAPTER TWO / OUR ESSENTIAL GOD 19
    Divine Law 21
    Illusion of Evolution 23
    Uniformitarianism vs. Catastrophism 28
    The Testimony of Scripture 29

CHAPTER THREE / THE POWER OF GOD 35
    A God of Miracles 35
    The Power of Man 36
    The Divine Mind 38
    Divine Faith 40
    The Spoken Word of Faith 42
    The Rebel 44
    Men of Power 46

PART TWO: THE EARTH
CHAPTER FOUR / WHY THE EARTH 53
    Earth's Measure of Creation 53
    In Their Steps 55
    A Second Birth 56
    A Living Earth 57
    "We Will Prove Them Herewith" 59
    Earthly Opposition 62
    The Worthy Dead 65
    The Destiny of Children 67
    An Eternal Inheritance 72

CHAPTER FIVE / EARTH'S DUAL CREATION 79
THE SPIRIT EARTH 79
    The Place of Death 81
    Paradise and Outer Darkness 83

Table of Contents

  Vision of the Spirit World 84
Creation of the Physical Earth 89
  The Gods of Creation 89
  Matter Unorganized 91
  Place of Creation 94
  The Two Accounts 95
  Dual Creation Accounts? 97
  Age of Earth 102
  The Time-frame of Creation 104
  The Six Times of Creation 107
  The Creation Days 107
  Summary of Earth's Organization 109
  The Second Council 112
  The Seventh Time/Day 113
Sequence of Creation 115
  Procreation of Man 116
  Earth's Two Sabbaths 117
  The Earth's Dual Weeks (Chart) 120

Chapter Six / The Fall of Earth 131
  Edenic Conditions 131
  Fall of Man 133
  The Garden of Eden 135
The Natural Effects of the Fall 136
  Fall from Celestial Time 136
  Fall from Primeval Light 137
  Fall from Primeval Place 138
  Fall from Immortality 139
  The Fall of Nature 140
Subsequent Curses and Falls 142
  Cain 143
  The People of Canaan 144
  Enoch 145
  Methuselah 145
  Sodom and Gomorrah 146
  Famine 146
  A Testimony to Sin 148
  Our Aging Earth 150

Chapter Seven / The Baptism of Earth 157
  Satan and the Serpent 158
  Wickedness of Men 160
The Global Flood 162
  The Ark 162
  The Rains 162

Whence the Waters 164
Reality of Flood 165
Sequence of the Flood 168
Ancient Mountains 169
The Flood was Justified 172
Earth's Baptismal Covenant 173
The Sign of an Oath 175
The Aftermath 176

CHAPTER EIGHT / EARTH IS DIVIDED 185
The Primal Land 185
ENOCH'S CITY OF HOLINESS 186
Post-Earth Ministry of Enoch 190
Melchizedek and His People 192
THE GREAT DIVISION 192
Pangaea and Continental Shift 194
THE LOST TRIBES OF ISRAEL 196
Israel's Beginnings 196
The Samaritans 199
A Remnant Led Away 200
Jehovah's Other Sheep 203
VIEWS OF LOCATION OF LOST TRIBES 204
Polar Regions 205
Scattered Among the Nations 208
A Single Body 209
Beyond the Earth 210
Conclusion 216

CHAPTER NINE / EARTH IN THE MERIDIAN OF TIME 223
SIGNS OF JESUS' BIRTH 223
Universal Light 223
A New Star 224
SIGNS OF JESUS' DEATH 225
The Redeemer Suffers 225
The Crucifixion—Earth Mourns 226
Darkness Covers the Earth 227
The Veil of the Temple 228
Judgment on American Israel 229
The Scandal of Eternity 231
Creation Travails 233

CHAPTER TEN / EARTH IN LATTER DAYS 237
Withdrawal of the Spirit 238
Man-Made Judgments 241
THE LORD'S SERMONS 243
The Wrath of God 243

Table of Contents

Zion and the Lord's Scourge	245
THE RETURN OF THE LOST TRIBES	250
Keys for Leading Lost Tribes	251
John the Revelator	252
The Return	254
Enemies A Prey	258
Time of Return	259
Rich Treasures	260
THE RETURN OF ENOCH'S ZION	262
Zion Above, Zion Beneath	262
Zion Expands	263
Sign of the Son of Man	264
SIGNS ON EARTH AND IN HEAVEN	266
The Half Hour of Silence	266
The Vanished Rainbow	267
Stars Shall Fall	267
The Earth Shall Reel	269
CHAPTER ELEVEN / EARTH'S DAY OF REST	277
CHRIST COMES IN GLORY	278
The Great Unveiling	278
Earth's Spiritual Baptism	279
A Paradisiacal Heaven and Earth	282
THE RESTORATION OF ALL THINGS	283
Reunion of the Continents	284
The Great Deep	286
The Mountains are Humbled	287
Magnitude of the Millennial-Earth	288
Earth is Healed	289
Universal Peace and Justice	292
Earth's Hour	296
CHAPTER TWELVE / THE CELESTIAL EARTH	303
The Little Season	303
The Last Resurrection	304
The Death of Earth	305
Resurrection of Earth	306
The Presence of the Father	308
A Sea of Glass	309
Powers of Celestial Beings	311
Two Celestial Cities	312
The God of Earth	313
The Meek Shall Inherit the Earth	314
Epilogue	317
Index	325

Introduction

Joseph Smith was instrumental not only in the restoration of ancient "plain and precious things" but also in bringing forth "things which never had been revealed from the foundation of the world" (D&C 128:19). Included in this latter-day spiritual treasure is astonishing information concerning the planet Earth and its prominent role in God's plan of salvation. We now know with greater understanding that Earth and mankind are far more than passing incidents in the course of cosmic happenings. Neither is a product of blind chance, and neither is destined for eventual oblivion. They are meant to abide forever in a symbiotic relationship binding each to each in the eternities to come.

This is a scriptural and prophetic account—both ancient and modern—of Earth's role in that relationship. As such, it presents some views admittedly at variance with current scientific thought. Science, being a human enterprise necessarily bound by empiricism, has but one arena in which to operate: natural law. But by definition revealed religion is not limited to the prevailing physical order, or to the assumption that God's works and ways must be circumscribed within it. For even though he is a God of law, he is not bound by law as we experience it. To so limit him would be a logical fallacy.

Authentic revelation is unbounded; it transcends time, space and circumstance. It speaks from the mountain top while science speaks from the valley. Different vantage points make for different points of view. For the present, their somewhat diverse claims must be examined and judged independently. However, both the religionists and the scientist must be prepared to abandon every error and misconception without regard to its author—be he prophet or savant. "True religion," said President Ezra Taft Benson, "accepts and embraces all truth; science is slowly expanding her arms and reaching into the invisible domain, in search of truth. The two are meeting daily; science as a child; true religion as the mother."[1] In due course, they will stand side by side on the summits of authentic knowledge and see eye to eye.

This Eternal Earth is written from the viewpoint of a scriptural conservative. Those of this persuasion respect the fundamental integrity of Holy Writ. While recognizing the presence of some myth, symbolism and hyperbole,[2] they feel obliged to interpret most of it—including the miraculous—quite literally. Hence, they agree with Joseph Smith: "the scriptures say what they mean and mean what they say."[3] Another conservative was Orson Pratt, a leading Mormon doctrinarian:

> We do not read the Scriptures as most of the inhabitants of the earth do, thinking that they must be spiritualized. There are scarcely any of the prophecies but what this generation, as well as some of the past generations, interpret them as meaning something altogether different from the reading of them. They look upon inspired men as saying one thing and meaning another, and the only way to ascertain what meaning they really wish to convey is to get an uninspired man to give some other meaning entirely different from the literal construction of the words of the inspired writer (JD, 20:8-9).

Elder Pratt went on to say when Latter-day Saints received the Holy Ghost, they learned that Scripture was to be understood "in the same light, and in the same sense" as they would understand any uninspired document.

While Scripture is open to varying interpretations, all are not of equal validity. Joseph Smith remarked, "We may spiritualize and express opinions to all eternity but that is no authority."[4] Any interpretation should be objective, logical and consistent. Proof-texting, the violation of context, is a widespread, and misleading practice; so is the unwarranted imposition of figurative or subjective meaning. Ideally, the same inspiration which produced a given scripture should be present when it is read. While human reason and research are legitimate tools in the quest for divine truth, they are not the only tools, nor even the best tools. "The best way to obtain truth and wisdom," said the Prophet, "is not to ask it from books, but to go to God in prayer, and obtain divine teaching."[5] However, very few of us are qualified to do so.

Holy Writ, including Joseph Smith's revelations, have been augmented by his own relevant statements as well as those of Brigham Young, Orson Pratt and other responsible Church authorities who have written or spoken on the subject at hand. This is done with the understanding that the LDS Church has no official position on some of their views or certain concepts treated herein.

Brigham Young, second president of the Church, described himself as "an apostle of Joseph Smith"[6] and looked upon the Prophet as his spiritual mentor:

> On the things of God, on the building up of his kingdom, on the doctrines Joseph taught, or on anything that pertains to the priesthood, I will not lay my memory aside to prefer any man's living. I know how I received the knowledge that I have got. I have seen the time when I first saw Joseph that I had but one prayer and I offered that all the time. And that was that I might be permitted to hear Joseph speak on doctrine and see his mind reach out untrammeled to grasp the deep things of God. But in consequence of the wickedness of the children of men and the consequent inability they possessed to receive heavenly things, he could not impart what was made known to him of the Lord. I was with him several years before I pretended to open my mouth to speak at all, but I would constantly watch him, and if possible learn doctrine and principle beyond that which he expressed. . . .An angel never watched him closer than I did, and that is what has given me the knowledge I have today. I treasure it up and ask the Father in the name of Jesus to help my memory when

information is wanted and I have never been at a loss to know what to do concerning the Kingdom of God.[7]

The Prophet testified: "I never told you I was perfect; but there is no error in the revelations which I have taught."[8] This statement is the underlying premise of this book; the propositions advanced and the conclusions reached are based primarily upon the validity and reliabilty of Joseph Smith's teachings and revelations. However, since it is possible for erroneous ideas to be extrapolated from true doctrine, it is not assumed that necessarily everything he said, or others claim he said, is factually correct. This applies even more to others quoted herein.

Mormonism is a faith rich in miracles, things which defy natural law. We will be considering a number of those involving the career of Earth in time and eternity. How we regard those claimed for the past will unavoidably effect our attitude toward those predicted for the future. For if we deny the one, can we affirm the other? I echo the Prophet Joseph's remark made eleven days before his martyrdom:

> I want to see truth in all its bearings and hug it to my bosom. I believe all that God ever revealed, and I never hear of a man being damned for believing too much; but they are damned for unbelief.[9]

Fallible men do not write infallible books. It is hoped, therefore, that the reader will look beyond the writer's errors and limitations to the wonder and magnitude of the story of this very unique and highly favored planet. Be assured, in due time a definitive biography will be forthcoming from its Creator.

Notes

1 *Teachings of Ezra Taft Benson*, 118. Orson Pratt said, "Science and true religion never can possibly contradict each other" (JD, 20:74)
2 Certain portions of Scripture are obviously of a symbolic or metaphorical nature and are not intended to be interpreted literally. Job, Ezekiel, Daniel and Revelation contain examples of this type of material.
3 TPJS, 264; see JD, 18:38.
4 WJS, 186.
5 TPJS, 191.
6 See JD, 3:212.
7 Discourse, 8 Oct. 1866, Church Archives.
8 TPJS, 368.
9 TPJS, 374.

Abbreviations

AF	Articles of Faith
CR	Conference Report
D&C	Doctrine And Covenants
DHC	Documentary History of the Church
DS	Doctrines of Salvation
ER	Evidences and Reconciliations
GD	Gospel Doctrine
JD	Journal of Discourses
MD	Mormon Doctrine
MS	Millennial Star
T&S	Times & Seasons
TPJS	Teachings of the Prophet Joseph Smith
VW	The Voice of Warning
KST	Key to the Science of Theology
WJS	The Words of Joseph Smith

Publisher's Note: All Italics are the author's emphasis unless otherwise indicated.

Part One

God and His Creations

Chapter One

The Universes

The Immensity of Creation

There are places on Earth where the night sky appears to be one vast canopy of lights—moving, yet motionless, remote, yet at our fingertips. It seems as though these far-flung worlds are virtually countless in number. However, reason reminds us that on the clearest and darkest of nights the unaided human eye can at best see only a few thousand heavenly bodies. Still, the emotional impression is closer to the ultimate truth than the visual fact—there are millions, billions, yes, *trillions* of spheres soaring through the known universe in silent splendor.

Not until the invention of the telescope did astronomers begin to grasp the magnitude of creation,[1] yet over five thousand years ago the patriarch Enoch testified: "And were it possible that man could number the particles of the earth, yea, *millions of earths* like this, it would not be a beginning to the number of thy creations; and thy curtains are stretched out still" (Moses 7:30). Abraham also learned—literally from the hand of the Lord—that his creations stretch on into the mists of endless time:

> And he said unto me: My son, my son (and his hand was stretched out), behold I will show you all these. And he put his hand upon mine eyes, and I saw those things which his hands had made, which were many; and they multiplied before mine eyes and I could not see the end thereof (Abr. 3:12).

There are countless other worlds inhabited by the family of Man located somewhere in or beyond the known cosmos. In the third of a series of three visions given him before he returned to Egypt, Moses "beheld many lands; and each land was called *earth,* and there were *inhabitants* on the face thereof" (Moses 1:29). The context of this passage makes it very clear that these "lands" were not islands or continents on this planet, but other worlds like Earth. Following this vision, the Lord proceeded to give Moses "only an account of *this earth and the inhabitants thereof"* (Moses 1:35).[2] But there is so much more!

The Galaxy[3] is about one hundred thousand light years in diameter.[4] Literally everything in it revolves around a dense cluster of stars comprising its nucleus. With its two satellite galaxies, the Magellanic Clouds,[5] it consists of well over a hundred billion stars, together with an unknown number of solar systems and enormous quantities of interstellar gas and dust—raw material for new stars and star systems.[6]

The stars, like so many electrons, appear to be in random orbits, but they are, in fact, under the direction of an overseeing Intelligence. Invisible hands direct this stellar traffic down the trackless highways of galactic space. The galaxies, with their teeming stars and planetary systems constitute wheels within wheels of time and motion. All is in flight; yet all is kept within the bounds and condition set by the Architect of Creation:

> And again, verily I say unto you, he hath given a law unto all things, by which they move in their times and their seasons;
> And their courses are fixed, even the courses of the heavens and the earth, which comprehend the earth and all the planets.
> And they give light to each other in their times and in their seasons, in their minutes, in their hours, in their days, in their weeks, in their months, in their years—all these are one year with God, but not with man" (D&C 88:42-44).[7]

The Sun has a mass three hundred thousand times that of Earth. But, as with people, so with planets—appearances can be deceiving. Earth, though fallen and presently dependent on the Sun, is in reality destined to shine forth with a glory the Sun will never know, even though it is now a rather small planet in a corner of a very large galaxy. Still, it is altogether unique among the nine planets of this solar system. Indeed, there are astrobiologists who maintain it may be unlike any other planet in the entire universe! For it alone seems to be life-creating and life-sustaining as we know life.

According to cosmologists, time began with the birth of the universe in the "Big Bang"—the initial explosion of incredibly compressed matter which set all in motion. Since that instant the universe is said to be expanding at almost unbelievable velocities as the pull of intergalactic gravity weakens and the galaxies speed outward and away from each other ever more rapidly.

We are learning that the universe is not only much larger, but also much more complex, than was previously supposed. Astrophysicists using increasingly sophisticated instruments, including high energy telescopes which operate from satellites, (such as the Hubble Space Telescope) have discovered or postulated such exotic objects as supernova, quasars, pulsars, neutron stars, black holes, antimatter, and dark matter.[8]

The universe is a strange and wonderful work. How did it come to be? And when? And by what or by whom? Is it purposeful or purposeless? No one this side of heaven can answer any of these questions with certitude. There are theories based upon present observations, but that is all they are—theories, working hypotheses. Distilled to their essence, they are to the effect that the universe is simply a product of physical law. Whether or not there was a Mind behind that law, a Creator behind creation, is not by its very nature a question amenable to scientific consideration. One thing is certain: there can be no purpose without a Purposer. If there was no Purposer behind the universe, then it is nothing more than a random, self-begotten accident, a product of meaningless chance. And if the universe is devoid of lasting meaning, then life itself is written on the wind.

Three Cosmogonies

In this regard, cosmologists provide us with scant comfort. Until this century it was assumed that the universe had existed, and would exist, forever. However in the last seventy-five years cosmologists have found evidence suggesting that all of the present star systems are doomed to eventual extinction.

Three main theories on the origin and fate of the universe were advanced in the twentieth century. The almost universally accepted present theory is known as the Big Bang. About fifteen billion years ago (estimates vary) matter was concentrated in a single mass of incomprehensible density with internal temperatures in the trillions of degrees. When this mass, the "cosmic egg," became critical, it exploded, hurling its minute particles of matter into space which thereafter under the force of gravity lumped together over billions of years to form the stars and galaxies. As the universe continues to expand, the galaxies, in moving ever farther apart, will cause intergalactic space to become increasingly empty. The primal hydrogen out of which the universe was formed will become essentially exhausted in each galaxy so that no new stars, the energy source of all life, will be created, and one by one the old stars will grow dim until, at last, all will be darkness and featureless emptiness.

In 1948 the steady state theory was put forth in opposition to the Big Bang. It posited a relatively stable but expanding universe, without beginning or end, in which the average density of matter is constant because hydrogen (the chief component of nuclear fusion) is being continually produced out of nothing.[9]

This *ex nihilo* hydrogen is used to form new stars which replace those that die, and also to fill in the elementary voids left by the galaxies as they continue to move outward and away from their primal center and each other. For a number of reasons, the steady state is now largely rejected.[10]

The third theory—the oscillating universe—was something of a synthesis of the steady-state and Big Bang cosmogonies. It held that the universe is going through endless cycles of contraction and expansion. Now expanding, gravitational attraction eventually will bring the galaxies to a halt; they will then reverse direction and contract once more into a single mass. When this mass reaches a critical density and temperature, a new Big Bang will occur and a new hydrogen universe will be born.[11]

Which, if any, of these cosmogonies is true? Did the universe explode into being only to eventually suffer eventual extinction as suggested by the Big Bang? Can new hydrogen be created out of nothing as required by the steady state? Or is it a self-perpetuating system, endlessly dying and reincarnating itself in an oscillating universe?[12] All three theories involve extensive observation and data, along with very esoteric mathematics. They are complex, tentative, and unverified. Obviously, if the truth were known there wouldn't be *three* theories, or *any*! Further research will doubtlessly see the present prevailing theory refined, modified, or discarded. A thoughtful scientist philosophized:

> The vast extent of the universe, both in space and in time, is, from the human point of view, completely aimless. Those immense clumps of matter, in their millions of millions, incessantly pouring out an inconceivably furious energy for millions of millions of years seem to be completely pointless. For a fleeting moment man has been permitted to stare at this gigantic and meaningless display. Long before the process comes to an end, man will have vanished from the scene, and the rest of the performance will take place in the unthinkable

night of the absence of all consciousness. This revelation is startling. It is still more startling, almost incredible, when we reflect that this amazing panorama sprang suddenly into existence a finite time ago. It emerged full-armed, as it were, *out of nothing*, apparently for the sole purpose of blazing its way to an eternal death. This is the scientific account. It seems to be true as far as it goes, but we cannot believe it is the whole truth.[13]

Is the universe a magnificent, but "meaningless display"? Is it a product of random chance? If it was created, was it by a deistic god manifesting himself in one capricious burst of creativity, only to watch thereafter as evolution took its glacial time producing life in all of its myriad forms from the initial hydrogen out of which all things supposedly evolved only to end in meaninglessness?

Is this the "whole truth"? The Second Law of Thermodynamics—with its mandate of change, decay, and death—is limited to the known universe. But, according to Joseph Smith, such a law does not exist in that universe where celestial law prevails. Is there such a universe? Human science stands mute before the things of eternity. It is no more capable of proving or disproving the existence of an immortal universe than it is of proving or disproving the existence of immortal men. But if there are immortal men, can we doubt the existence of immortal worlds whereon they dwell? We have come to the crossroads of knowledge. It is precisely at this point that the science of Earth should yield humbly to the science of Heaven.

The Eternal Universe

As modern astronomy unlocked the immensity of the temporal order of creation, so did the Restoration unlock the immensity of the eternal order of Gods, men, and worlds. Brigham Young declared:

> How many Gods there are, I do not know. But there never was a time when there were not Gods and worlds, and when men were not passing through the same ordeals that we are now passing through. That course has been from all eternity, and it is and will be to all eternity. You cannot comprehend this; but when you can, it will be to you a matter of great consolation (JD, 7:333).

The previously quoted words of Enoch, Abraham and Moses lend credence to Brigham Young's statement. It is also supported by one of the last and most profound doctrines revealed by Joseph Smith: "Intelligences exists one above another, so that there is *no end* to them. . . .If Jesus Christ was the Son of God, and John discovered that God the Father of Jesus Christ had a Father, you may suppose that *He* had a Father also. Where was there ever a son without a father? And where was there ever a father without first being a son?" (TPJS, 373; emphasis in the original).[14] The virtual infinity of organized life and worlds on which they dwell is reflected in these lines by W.W. Phelps:

> If you could hie to Kolob
> In the twinkling of an eye,
> And then continue onward,
> With that same speed to fly,
> Do you think that you could ever,
> Through all eternity,
> Find out the generation
> Where Gods began to be?
> *(LDS Hymnal, 284)*

That our finite minds cannot comprehend the revelation that organized life is literally without beginning or end is at once both understandable and irrelevant. The fact remains, the Gods and their works are endlessly on-going realities in this and other eternities. This is implicit in the words of the Creator to Moses:

> The heavens, they are many, and they cannot be numbered unto man; but they are numbered unto me, for they are mine. And as one earth shall pass away, and the heavens thereof even so shall another come; and there is no end to my works, neither to my words (Moses 1:37-38).[15]

If there is "no end" to God's labors, can there be a beginning? Not according to Joseph Smith:

> That which has a beginning will surely have an end; take a ring, it is without beginning or end—cut it for a beginning place and at the same time you have an ending place (TPJS, 181; see 354).

Just as all men are not mortal—some being spirits, some translated, and some resurrected—so, too, some worlds are formed of spirit matter, some of gross matter, and some a combination of the two.[16] Thus, planets, stars, solar systems, galaxies—perhaps even universes—are organized under different laws, of different materials, at different times, and for different purposes. For instance, the revealed purposes of the temporal sun, moon, and stars are to "divide the day from the night," to be "for signs, and for seasons, and for days, and for years" and for "lights in the firmament of the heaven to give light upon the earth" (Moses 2:14-15; see Gen. 1:14-15). Note, they are not said to be for habitation. These spheres are Earth-centered; they serve the eternal purposes of God vis-a-vis mankind on this and like worlds. And the galaxies, together with the cosmic matter of intergalactic space, constitute vast reservoirs of material from which new spheres may be organized in the eons to come.

Prior to learning the actual composition of known heavenly bodies, it was thought that at least some of them were inhabited. For the 17th century German astronomer, Johannes Kepler, distance lent enchantment. He so admired the Sun that he considered it "worthy of the Most High God, should he be pleased with a material domicile and choose a place in which to dwell with the blessed angels" (Sullivan, 129). Had Kepler known that the surface temperature of the seething Sun is about 11,000 degrees Fahrenheit[17] and that, like all such stars, the Sun is riven with continual upheavals, he might have thought it better suited to hell than to heaven. However the physical composition of our gaseous Sun was yet to be understood, so Brigham Young, along with others, expressed an opinion:

> Who can tell us of the inhabitants of this little planet that shines of an evening, called the moon? . . .So it is with regard to the inhabitants of the sun. Do you think it is inhabited? I rather think it is. Do you think there is any life there? No question of it; it was not made in vain. It was made to give light to those who dwell upon it, and to other planets; and so will this earth when it is celestialized (JD, 13:271).[18]

Joseph Fielding Smith expressed a similar belief:

> *It is my opinion that the great stars that we see, including our sun, are celestial worlds; at least worlds that have passed on to their exaltation or other final resurrected status* (DS, 1:88-89; emphasis in original).[19]

However, Orson Pratt thought otherwise.

> I will not say as the splendor of our sun, for it is not a celestial body. Although the light of the sun is very glorious, it will not begin to compare with that of this earth, when it becomes celestial and eternal and is lightened by the presence of God the Father. It is doubtful whether the children of mortality on other worlds, will ever behold the light of this earth, after it is made eternal, unless they happen to catch a glimpse of it by vision. God dwells in a world of light too

glorious for mortal eyes to behold, unless aided by the Spirit of the living God (JD, 19:290-91; see 14:236-37).

The data of twentieth century astrophysicists, appear to preclude the likelihood of even immortal beings inhabiting (or wanting to inhabit) the Sun or any other known stars having similar characteristics. They write of the "violent storms that rage across the face of the sun in an 11-year cycle, geysers of hot gas that rise hundreds of thousands of miles into the solar atmosphere, and explosive outbursts of gamma rays, x-rays, ultraviolet radiation, and energetic protons that erupt sporadically from the surface."[20] Is heaven hidden beneath all of this infernal activity? The observed facts about the Sun and its peers make them poor candidates for the glorious realms of revelation.

Since this planet has a spiritual counterpart, may not the physical universe have one as well? From the standpoint of the gospel such a universe must exist. There is simply no place in the physical universe *as defined by cosmologists* for immortal, resurrected beings and immortal, resurrected worlds. However, modern revelation informs us that there is another *material* order, one immeasurably superior in glory, majesty, and perfection to all that we know.

This higher order may be considered the counterpart of this temporal order, one that is spiritual and, therefore, everlasting.[21] It is the universe of the Father's many mansions—the realms of immortal glory (see WJS, 370). In Joseph Smith's last published revelation they are described as *"the eternal worlds"*— the glorious habitations of immortal beings.[22] Discoursing on John's vision, the Prophet referred to immortal animals "saved from ten thousand times ten thousand earths like this"[23] —a hundred million resurrected earths!

Unlike the seething, gaseous cauldrons comprising the known mortal stars, the eternal worlds are enveloped in spiritual fire, the glory of God. The Prophet explained, "God almighty Himself dwells in eternal fire; flesh and blood cannot go there for all corruption is devoured by the fire. . . . All men who are immortal dwell in everlasting burnings."[24] He knew whereof he spoke. In 1832, the Prophet and Sidney Rigdon saw in vision the different glories men and women will inherit in the resurrection. Of this experience he wrote:

> But great and marvelous are the works of the Lord, and the mysteries of his kingdom which he showed unto us, which surpass all understanding in glory, and in might, and in dominion. . . .
> *Neither is man capable to make them known,* for they are *only to be seen* and understood by the power of the Holy Spirit, which God bestows upon those who love him, and purify themselves before him; To whom he grants this privilege of *seeing and knowing for themselves* (D&C 76:114-16).

A second vision in 1836, seen only by the Prophet, was limited to the highest glory:

> The heavens were opened upon us, and I beheld the celestial kingdom of God, and the glory thereof, whether in the body or out I cannot tell. I saw the transcendent beauty of the gate through which the heirs of that kingdom will enter, which was *like* unto circling flames of fire; Also the blazing throne of God, whereon was seated the Father and the Son (D&C 137:1-4).

Abraham, another great seer, also beheld and knew for himself:

> And I saw the stars, that they were very great, and that one of them was nearest unto the throne of God;[25] and there were many great ones which were near unto it; And the Lord said unto me: These are the governing ones; and the name of the great one is Kolob, because it is near unto me, for I am the Lord thy God: I have set this one to govern all those

which belong to the same order as that upon which thou standest (Abr. 3:2-3; see Moses 1:40).[26]

B. H. Roberts, reflecting the then-current view of the cosmos, wrote:

> This great star [Kolob] constitutes the grand center of our universe, around which all other stars revolve, or by which they are governed in their movements, or receive controlling force; and thence from this great central governing star, in ever widening circles revolve countless worlds and world-groups with their respective central suns in bewildering yet orderly magnificence and splendor (CHC, 2:28-29).

However, Abraham does not say that the governance of Kolob is universal. It is limited to a particular order of worlds—"all those *planets* which belong to the same order [of planets] as that upon which *thou standest*" (Abr. 3:9). Abraham stood upon a fallen, *inhabited* planet, not barren Mars, deadly Venus or some mortal star. Dead planets and volatile stars are not of the "same order" as this living world.[27] The "same order," therefore, appears to refer to that order of planets which, like Earth, are organized to become immortal realms when they "pass away." From such eternal stars as Kolob, planets like Earth are guided by divine beings toward their God-ordained destiny of future immortality.

Thus, Kolob bestows more than gravitational influence and cosmic energy on our Sun and Earth. It is a seat of heavenly authority and divine power. A "Grammar and Alphabet of the Egyptian Language" was prepared by Joseph Smith and Oliver Cowdery in connection with the translation of the papyri upon which the published writings of Abraham are presumably based.[28] It states:

> Kolob signifies the highest degree of power in government, pertaining to heavenly bodies. . . .the eldest of all the

Stars, the greatest body of the heavenly bodies that ever was discovered by man. . . .the first great grand governing fixed star which is the fartherest that ever has been discovered by the fathers which was discovered by [or revealed to] Methuselas and also by Abraham.[29]

How many other inhabited worlds does Kolob and its companions govern? Are they all to be found within the myriad star systems scattered throughout our Galaxy? It would seem likely, but we do not know.

Where is Kolob? It has been thought to be located at or near the physical center of the Galaxy. But is it? As before noted, so far as astronomers are concerned, all stars are mortal—they are born, grow old, and die. But Kolob is an immortal star with immortal inhabitants, does it dwell among the dying? Is any heaven to be found within the controlled chaos of the stars and galaxies described in every textbook on astronomy? Not that the stars are crammed together; they are isolated from one another by trillions of miles. However, the isolation of the temporal from the spiritual universe is not only spatial. Immortal worlds are *qualitatively* distinct from their temporal counterparts even as Gods and angels are distinct and apart from mortal men. From our perspective, the Kolobian cluster of redeemed worlds is located within the physical boundaries of the Galaxy. While from God's perspective, Kolob and his companions are infinitely apart in the splendor of an eternal universe.

When the Son of Man returns in all of his majesty, it will be asked: "Who is this that cometh down from God in heaven. . . yea, *from the regions which are not known*" (D&C 133:46; see Isa. 63:1). These are the regions of those immortal stars no earthly telescope, however powerful, will ever focus upon.[30] They soar nigh unto Kolob—that most glorious world *"nearest unto the throne of God."*

Notes

1. Galileo Galilei first used the newly invented telescope in 1609. Only in 1924 did Edwin Hubbel discover that, contrary to the then-prevailing view, our Milky Way was not the only galaxy. Present estimates of the number of galaxies in the universe range from fifty to several hundred billion.
2. Note that the term, "earth," is restricted to planets organized for, and inhabited by, man. Here is a perfect example of the superiority of revelation over reason. Although the Lord has spoken, modern science is not listening. It continues to debate if there is complex life anywhere else in the universe—not human life, but *any* life other than microorganisms!
3. When capitalized, "Galaxy" refers to this galaxy, the Milky Way.
4. The velocity of light is 186,283 miles per second, covering 5.87 trillion miles in a year. Most astronomical measurements are mathematical approximations.
5. They are known as "dwarf galaxies" since each consists of *only* 10 billion stars.
6. Stars are thought to be produced from great clouds of loose hydrogen atoms which first evolve into protostars and then into true stars. Their lifespan depends upon their size and structure and varies from as little as one million to more than a trillion years.
7. Orson Pratt defined "one year with God" as referring to "celestial time" (D&C, 1890 ed); see Abr. 3:4.
8. More than 90 per cent of the universe is believed to consist of dark matter—the supposed key to its structure and eventual fate.
9. J. Reuben Clark seemed to favor this theory apparently because it allowed for the death of some worlds while supporting the idea of an endless, dynamic universe. See, Clark, 53-61.
10. The steady state violates the accepted principle of the conservation of matter and energy (they can neither be created nor destroyed)—a view held by Joseph Smith: "The pure principles of element are principles which can never be destroyed; they may be organized and reorganized, but not destroyed. They had no beginning and can have no end" (TPJS, 351-52).
11. This scenario depends upon a sufficient density of matter and, hence, of gravitational attraction, to halt and reverse the outward movement of the galaxies. Lacking it, the lifeless universe will continue its voyage into space forever—the ultimate "Flying Dutchman."
12. It has been suggested that the steady state is compatible with Judaeo-Christian concepts, the oscillating Universe with Eastern religions, and the big bang with atheism.

13 J. W. N. Sullivan, 25-26.
14 Brigham Young said, "Brother Kimball quoted a saying of Joseph the Prophet, that he would not worship a God who had not a Father; and I do not know that he would if he had not a mother; the one would be as absurd as the other" (JD, 9:286).
15 See 2 Ne. 29:9. Referring to the organization of this earth, Joseph Smith said, "The grand councilors sat at the head in yonder heavens and contemplated the creation of the *worlds* which were created at the time" (TPJS, 348-49).
16 In Mormon thought, matter is divided into two basic classes: gross matter (called element) which fills the observable universe, and refined matter (spirit). See D&C 93:33-35. The purity of spirit matter precludes its detection by mortals who are composed of gross matter. Joseph Smith said, "There is no such thing as immaterial matter. All spirit is matter, but it is more fine or pure, and can only be discerned by purer eyes; we cannot see it; but when our bodies are purified we shall see that it is all matter" (D&C 131:7-8). As physicists involved in the study of quantum mechanics identify ever more exotic forms of matter, they seem to be approaching the veil separating the physical from the spiritual order of reality.
17 The Sun's corona reaches temperatures approaching one million degrees.
18 Johannes Kepler also believed the moon was inhabited.
19 Apparently his grandfather, Hyrum Smith, thought the Sun, moon and stars were inhabited (see Diary of George Laub, 34-35). In 1892, Oliver B. Huntington wrote "As far back as 1837, I know that he [Joseph Smith] said the moon was inhabited by men and women the same as this earth, and that they lived to a greater age than we do— that they live generally to near the age of a 1000 years." Huntington also claimed Joseph Smith, Sr. told him, "I should preach the Gospel . . . to the inhabitants of the moon. . . ." (The Young Woman's Journal, 3 (March, 1892), 263-64).
20 Jastrow & Thompson, 313.
21 The idea of other universes within or apart from the known universe is now postulated by cosmologists. Still, science does not contemplate the immortal worlds of scripture and revelation. Orson Pratt spoke of God forming "new worlds, new universes" (JD, 20:75).
22 See D&C 132:63; TPJS, 237, 306, 359; WJS, 232, 253, 354, 370.
23 TPJS, 291.
24 TPJS, 367; see 361,372-73..
25 Three stars, of different magnitudes, seem to represent the Godhead. The greatest is unnamed; it is simply "the celestial, or the residence of God," meaning, the Father. Kolob, "first" and "last," is "*nearest* to the

celestial" even as the Son, Alpha and Omega, is nearest the Father. Oliblish (representing the Holy Ghost?) is *"near"* the celestial also and, like Kolob, holds keys of power "pertaining to *other planets*" (see, Abraham, explanation of facsimile number two, figures 1 and 2).

26 In the Prophet's explanation of Facsimile two, Kolob is said to be "*nearest* to the celestial," while Oliblish, Kolob's companion, is described as being *"near* to the celestial. . .holding the key of power *also,* pertaining to *other planets."* "Nearest" and "near" may refer to their nature and glory vis-a-vis the celestial "residence of God" rather than to their physical distance from it.

27 However, Orson Pratt, limited by 19th century astronomy, assumed that the other planets in this solar system were, like Earth, inhabited by fallen humans. He identified these planets with the parable in D&C 88:51-61; see JD, 19:293-94).

28 The papyri was obtained by the Prophet in 1835. A portion was translated and published in 1842. The existence of the "Grammar" remained unknown until 1936 when it was discovered in the Church Archives by Dr. Sidney B. Sperry of Brigham Young University. Its reliability is a matter of debate among LDS scholars.

29 Quotes are taken from a photographic copy of "Grammar and Alphabet of the Egyptian Language" in writer's possession.

30 Paul declared Christ to be the Creator of all things "visible and *invisible"* (Col. 1:16). Worlds are immortal when they are spiritual rather than temporal in organization (see D&C 88:27). Being resurrected bodies, their original spirit organization has become "inseparably connected" or fused to its temporal counterpart. (See D&C 93:33; Alma 11:45.) This fusion of spirit with element produces a unified spiritual body, the nature of which is far different than those worlds of gross materiality found in this universe. Consequently, we cannot see them for spirit "can only be discerned by purer eyes" than mortals possess (see D&C 131:7-8).

Chapter Two

Our Essential God

Impressive as are mankind's achievements in the many fields of human endeavor, they need to be put in perspective. Moses was raised in the luxury of Pharaoh's court, and was familiar with the magnificence of ancient Egypt. Thereafter he had several astonishing visions, the first of which was of this solitary planet and its inhabitants. Overwhelmed by what he was shown, upon recovering he thought to himself, "Now, for this cause I know that man is nothing, which thing I never had supposed" (Moses 1:10). Of course Moses was speaking relatively. The point is, compared to the works of God, we have good cause to be humble. Elder Bruce R. McConkie wrote:

> We know but little about God and his glory and the laws by which he created worlds without number. What do we know about the sidereal heavens and the endless galaxies spinning through boundless space? Our knowledge of the creation of even this little dot of dust known as planet earth is so slight that we can scarcely envision its place in the eternal orbits and its relationship to the endless worlds that roll everlastingly forth from the Creator's hand.[1]

Left to ourselves, we only know what is knowable through our finite senses and their technological extensions. Consequently, the working philosophy of modern science is necessarily naturalism—the assumption that reality is limited to the physical universe; there are no spiritual or supernatural realities.[2] Elder John A. Widtsoe wrote, "The gospel accepts God as the author of all knowledge; science gathers facts and tries to interpret them without reference to a Supreme Being" (ER, 126). While many scientists believe in God and are religiously oriented, others go about their business as though he was either non-existent or irrelevant—honoring the art but ignoring the Artist. As for Holy Writ, their basic attitude seems to be that where reality is concerned, Scripture has nothing to say and no reason to say it. Granted, the Christian church of the middle ages is rightly faulted for its hostility toward free inquiry, but a greater error is made when blind reason supplants blind faith. The wedding of science to naturalism will prove to be a marriage of convenience rather than necessity. Something of a mathematician himself, Orson Pratt testified of spiritual realities and higher laws:

> Now, the Lord has powers beyond those with which we are acquainted. He has almighty powers. He has only intrusted us his children of mortality with a knowledge of some of the more *gross principles and laws of this fallen creation* and when we, through hard study, search out the relation of one law to another, we think we are learned men; but I think when we learn in that great university the sciences of which the Lord our God is the great Teacher, we shall learn more rapidly and comprehend more easily the things of his kingdom, than we now do the things of time (JD 19:294; emphasis added).

Unfortunately, some "learned men" deny the existence of higher laws and, therefore, of a higher Power. Pure mechanists maintain that the universe was produced solely by four factors: weak and strong nuclear force, electromagnetism, and gravitation. Theists, acknowledging the Creator, maintain that he *discovered* those physical principles which operate always and everywhere. They view him as a cosmic engineer who, in the process of coming upon self-existing physical laws, and by obedience to them, produced the universe—a seemingly serendipitous achievement.

This approach is appealing because it salvages God while providing man with a sense of stability and control where law is concerned. However, God would be circumscribed by the laws he has discovered; he would be as other men. He could not rightly claim to be omnipotent because he could not be certain he was omniscient.[3] As Elder Joseph Fielding Smith asked, *"will the Almighty find some hidden truth or law which will shatter all?"*[4] If law is supreme, the Creator would be obliged to abdicate in favor of his own handiwork. In enthroning law, theists, like mechanists, unwittingly dethrone the Law-giver.

Divine Law

To all intents and purposes, God created law. However, since there have always been Gods, there have always been laws. God and law, like God and Priesthood, are essentially co-eternal. Orson Pratt remarked:

> I have no doubt in my own mind but what there have been laws from all eternity—or if you do not wish to call them laws, call them forces, call them powers. . .and then again these laws or forces have also been under the control of a wise, supreme intelligence from all eternity to the present time (JD, 20:74).

This statement must be tempered by another of his comments: "Substance had no beginning; to say that laws had no beginning would be another thing; some laws might have been eternal, while others might have had a lawgiver" (JD, 19:286).

The eternities are organized into various kingdoms or systems of law with their appropriate bounds and conditions.[5] "All kingdoms have a law *given*. . . . And unto every kingdom is *given* a law; and unto every law there are certain bounds and conditions" (D&C 88:36, 38). Joseph Smith taught: "God has made certain decrees which are fixed and immovable; for instance,—God set the sun, the moon, and the stars in heaven, and gave them their laws, conditions, and bounds which they cannot pass, except by his commandments" (TPJS, 197-98). Orson Pratt agreed: "God is the great Author of all law, and is just as able to counteract a law, as he is to continue a law" (JD, 21:238).[6]

President John Taylor asked: "Who has implanted certain principles in matter and in all creation? God has done it. All things are subject to these laws" (JD, 25:215).

The organizing, governing, enlivening principle in all things—the substance permeating all things, animate and inanimate, is in actuality, the Spirit of God,[7] the light of Christ:

> Which light proceedeth from the presence of God to fill the immensity of space—The light which is in all things, which giveth *life* to all things, which is the *law* by which all things are governed, even the *power* of God (D&C 88:12-13).

As the spirit is the life of the body, so is the light of Christ the life-principle in *all* things. By virtue of this spiritual *substance* which fills the immensity of space, including the voids between the stars and galaxies, the Almighty creates, governs, sustains, and redeems all things temporal and spiritual. This divine light or spiritual force is the irreducible foundation of all things, whether spirit or element, whether animate or inanimate, whether sub-atomic particles or galaxies. Being the "power of God," the light of Christ is tantamount to the power of the Holy Priesthood, which John Taylor defined as, "the government of God, whether on the earth or in the heavens, for it is by that power, agency, or principle that all things are governed on the earth and in the heavens, and by that power that all things are upheld and sustained" (*GK*, 129). God, Spirit, Priesthood—

these are indivisible and supreme over all physical and spiritual law. Law is the servant, not the master, of the Gods in eternity.

It is those inseparable twins, pride and ignorance, which prompt men to deny or limit the powers of the Creator. As Joseph Smith, observed: "It is the constitutional disposition of mankind to set up stakes and set bounds to the works and ways of the Almighty." He added, "I say to all those disposed to set up stakes for the Almighty, you will come short of the glory of God" (TPJS, 320-21). But while this disposition is to be expected in the world at large, it would be inappropriate and disingenuous were such a mind-set to exist among those professing faith in, and committed by covenant to, the Son of God.

Learning by reason and experience should never be at the expense of learning by valid faith and true revelation which bow to the superior wisdom of the Creator. Christ warned: "And in nothing doth man offend God, or against none is his wrath kindled, save those who confess not his hand in all things" (D&C 59:21). No worthy enterprise justifies the absence of humility—the recognition of God's works and ways and of our fundamental dependence upon him.

Illusion of Evolution

Belief in the supremecy of natural law explains the pervasiveness of evolutionary doctrine. Evolution has become the new god of creation for those who view it as the directing principle behind virtually everything. Sir Julian Huxley, a rabid evolutionist, so viewed it:

> The concept of evolution was soon extended into other than biological fields. Inorganic subjects such as the life-histories of stars and the formation of the chemical elements on the one hand, and on the other hand subjects like linguistics, social anthropology, and comparative law and religion, began to be studied from an evolutionary angle, until today we are

enabled to see evolution as a universal and all-pervading process.[8]

It is well to recall Elder John A. Widtsoe's caveat: "Science is trustworthy as far as human senses and reason are trustworthy—no more" (*ER*, 133). While no one should be faulted for advancing any hypothesis they choose, they can be justifiably criticized for practicing sleight-of-hand and transforming questionable theory into unquestionable fact. Elder Widtsoe, a scientist himself, pointed out:

> . . .the interpretations of observed facts must be distinctly labeled as *inferences, and not confused with facts*. . . .There must be a distinct segregation of facts and inferences in the utterances of scientific men (*ER*, 127-28; emphasis added).

He also noted:

> The man, learned or unlearned, who declares the doctrine of the common origin of life on earth to be demonstrated beyond doubt, has yet to master the philosophy of science. The failure to differentiate between facts and inferences is the most grievous and the *most common sin of scientists* (*ER*, 153; emphasis added).

There are very solid scientific challenges to organic evolution on the *macro level*.[9] Indeed, no other hypothesis would be seriously entertained unless comparable objections were first resolved. Fifty years ago highly controlled experiments by Miller, Urey and others were conducted to simulate presumed conditions in the earth's early atmosphere. Their purpose was to determine if supposedly very simple life forms could have evolved from inorganic matter—abiogenesis. The results, partially successful in producing a number of simple amino acids, were far from conclusive. However, writes Dean L. Overman, "Random abiogenesis became the accepted theory in college textbooks despite the absence of evidence supporting this view."[10] It is still taught as fact in spite of increasing challenges by reputable scientists. We now know that the simplest cells are incredibly complex. Sir Fred Hoyle and Chandra Wickramsinghe "concluded that life could not have appeared by earthbound random processes even if the whole universe consisted of primeval soup." Wickramsinghe succinctly summed up the matter: "The chances that life just occurred are about as unlikely as a typhoon blowing through a junkyard and constructing a Boeing 747" (Overman, 59-60). The origin of life, even of microorganisms, is as much a mystery as ever.

Why has evolutionary theory received such wide acceptance? Do men want to replace the Creator? Is it because evolution is "pleasing unto the carnal mind"? Is it because it liberates the conscience by justifying mankind in behaving like animals since they are animals? One thing is certain: no other supposedly scientific discovery serves the father of all lies so well as this morally and spiritually pernicious doctrine.

And it originated with men who were contemptuous of Scripture. Hugh Nibley writes of Lamarck's and Darwin's antipathy toward the Bible:

> The great Lamarck, before he even came up with his explanation of the creation, was animated "by a severe . . . philosophical hostility, amounting to hatred, for the tradition of the Deluge and the Biblical creation story, indeed for everything which recalled the Christian theory of nature." And Darwin writes of himself in his twenties: "I had gradually come, by this time, to see that the Old Testament from its manifestly false history of the world and from its attributing to God the feelings of a revengeful tyrant, was no more to be trusted than the sacred books of the Hindoos [sic], or the beliefs of any barbarian. . . . By further reflecting. . . that the more we know of the *fixed laws of nature the more incredible do miracles become*—that the men at that time were ignorant and credulous to a degree almost incomprehensible to us

....This disbelief crept over me at a very slow rate, but was at last complete. The rate was so slow that I felt no distress, and have never since doubted for a single second that my conclusion was correct.

Professor Nibley then makes a telling observation:

Students commonly assume that it was the gradual amassing of evidence that in time constrained such men to part company with the Bible. Exactly the opposite is the case: long before they had the evidence, they brought to their researches such an unshakable determination to discredit the book of Genesis that the discovery of the evidence was a foregone conclusion. . . .*The battle was against revelation, and evolution was the weapon forged for the conflict.*[11]

With blithe candor, Huxley acknowledged what theistic evolutionists will not: there is no place for the God of Scripture in the theory of organic evolution—whether vintage Darwinism with its emphasis on "natural selection," or neo-Darwinism, with its faith in "beneficial mutations":

Darwinism removed the *whole idea of God* as the creator of organisms from the sphere of rational discussion. Darwin pointed out *that no supernatural designer was needed;* since natural selection could account for any known form of life, there was *no room for a supernatural agency* in its evolution....I think we can *dismiss entirely all idea of a supernatural overriding mind* being responsible for the evolutionary process (Morris, 25; emphasis added).

But theistic evolutionists, wanting it both ways, argue that organic evolution was simply the method God used in creating life. However, the word of the Lord through his modern prophets is that, at least *where man is concerned,* it was not. To that end, in 1909 the First Presidency, responding to inquiries regarding evolution, issued an official doctrinal statement, "The Origin of Man:"

> "God created man in his own image, in the image of God created he him; male and female created he them." Moses. . .when making this important announcement, was not voicing mere opinion. . . .He was speaking as the mouthpiece of God, and his solemn declaration was for all time and for all people. . . .It is held by some that Adam was not the first man upon this earth, and that the original human being was a development from lower orders of the animal creation. These, however, are the theories of men. . . .Man began life as a human, being in the likeness of our heavenly Father. . . .The Church of Jesus Christ of Latter-day Saints, basing its belief on divine revelation, ancient and modern, proclaims man to be the direct and lineal offspring of Deity.[12]

The genealogical stream of mankind flows directly from divine beings to the human race with no lesser forms of life or hominids intermixed between them to debase it.

The apostle Paul asked, "Hath not God made foolish the wisdom of this world?" (1Cor.1:20). Although ridiculed and rejected, in time the revelations of Abraham, Moses and Joseph Smith pertaining to the origin of man will prove correct, and the arguments for the human race being a product of evolutionary processes will be viewed on a par with those advanced for a flat earth!

Why defend the mirror image of the truth? Earth did not create man, *Man created Earth.* Man did not evolve from inorganic matter or microbial organisms in the sea; he set foot upon a paradisiacal planet in perfection and *immortality.* No, the origin of the human race is not to be found in the Rift Valley of East Africa, but in the eternal worlds of the Gods. Even then, we face the question posed by W. W. Phelps:

> Do you think that you could ever,
> Through all eternity,
> Find out the generation
> Where Gods began to be?

Uniformitarianism vs. Catastrophism

Associated with the general theory of evolution is uniformitarianism—a working hypothesis of those geologists concerned with Earth's beginnings.[13] Such research poses a dilemma: how can we determine the exact nature of this planet in its primordial state when we cannot replicate that state? The answer provided by uniformitarianism is we must assume that nature's law-controlled processes are essentially constant and unchanging—the unknown past can be extrapolated from the known present. Understandably, present theory necessarily limits the Almighty to one method of creation and one system of law throughout the universe. While this premise is necessary, it should not be affirmed with rigid dogmatism. But as Peter foretold, uniformitarianism is the ruling gospel: "All things continue as they were from the beginning of the creation" (2 Pet. 3:4).

Opposing the doctrine of uniformitarianism is the Bible-centered doctrine of catastrophism, the belief that the earth has experienced divinely-initiated cataclysmic changes in the past, and will do so in the future. Catastrophism is projected in connection with the Creation, the Fall, the Flood, the Second Coming of Christ, and other miraculous events described in the Bible. Creationists feel that these extraordinary interruptions in the normal course of nature's operations serve to disprove uniformitarianism. There are reputable scientists among Creationists and others of like thinking.

Special creationists—those who defend the Biblical account of the creation of the earth—are at a great disadvantage because their scriptural arguments are almost wholly dependent upon Genesis. Latter-day Saints, however, possess a considerable body of new scripture and prophecy having a direct bearing on the subject which indicate that the Earth and man are far different from what they were when first organized. Earth is a fallen planet; mankind is a fallen race.

The Testimony of Scripture

For those who are "ever learning, and never able to come to a knowledge of the truth," all is nature, all is chance, all is purposelessness. However, the Prophet Joseph Smith, who beheld realities few men have been privileged to behold, knew better:

> But if this life is all, then why this constant toiling, why this continual warfare, and why this unceasing trouble? But this life is not all; the voice of *reason,* the language of *inspiration,* and the Spirit of the living God, our Creator, teaches us, as we hold the record of truth in our hands, that this is not the case, that this is not so; for, the heavens declare the glory of a God, and the firmament showeth His handiwork; and a moments reflection is sufficient to teach every man of common intelligence, that all these are not the mere productions of *chance,* nor could they be supported by any power less than an Almighty hand (TPJS, 56; emphasis in original).

Scripture repudiates the philosophy of naturalism. It sustains the cosmological argument for God in declaring him to be the Great First Cause.[14] The most philosophical statement on God's essentiality in scripture is made by the prophet Lehi who explains that in the absence of God, theoretically there could only be a meaningless void! He also advances the teleological argument[15] for God in these words:

> Wherefore, it [vacuous existence] must needs have been created for a thing of naught; wherefore there would have

been no purpose in the end to its creation. Wherefore, this thing [meaninglessness] must needs destroy the wisdom of God and his eternal purposes, and also the power, and the mercy, and the justice of God (2 Ne. 2:12).

Such abstractions as law, sin, righteousness, happiness, and misery would not exist. Lehi adds:

> And if these things are not there is no God. And if there is no God we are not, neither the earth; for there could have been no creation of things, neither to act nor to be acted upon; wherefore, all things must have vanished away (2 Ne. 2:13).

Thus, there could have been no Big Bang because there would have been absolutely nothing to produce it! In essence, Lehi maintains that in the absence of abstract moral law there can be no concrete physical law; in the absence of spiritual realities there can be no physical realities. All things are bi-conditional. And they exist only because God exists.

Referring to the Son of God, the apostle John testified: "All things were made by him; and without him was not anything made that was made" (John 1:3).[16] A second witness, the apostle Paul, wrote:

> For by him were all things created, that are in heaven, and that are in earth, visible and invisible, whether they be thrones, or dominions, or principalities, or powers: all things were created by him, and for him. And he is before all things, and by him all things consist (Col. 1:16-17).

Scripture is replete with such testimonies of the divine origin of creation.

Because the Lord is the source of all things temporal and spiritual, he provides both temporal and spiritual evidences of his existence. The pre-mortal Christ told Adam:

> And behold, all things have their likeness, and all things are created and made to bear record of me, both things which are temporal, and things which are spiritual; things which are in the heavens above, and things which are on the earth, and things which are in the earth, and things which are under the earth, both above and beneath: all things bear record of me (Moses 6:63).[17]

Truly, "The heavens declare the glory of God; and the firmament sheweth his handiwork" (Ps. 19:1). The Creator himself declared the visible universe a virtual proof that he lives:

> The earth rolls upon her wings, and the sun giveth his light by day, and the moon giveth her light by night, and the stars also give their light, as they roll upon their wings in their glory, in the midst of the power of God. . . .Behold, any man who has seen any or the least of these hath seen God moving in his majesty and power (D&C 88:47).

The avowed atheist, Korihor, tried to deny the undeniable God. He demanded a sign before he would believe. But Alma, employing cosmological argument, rebuked him:

> Thou hast had signs enough; will ye tempt your God? Will ye say, Show unto me a sign, when ye have the testimony of all these thy brethren, and also all the holy prophets? The scriptures are laid before thee, yea, all things denote there is a God yea, even the earth, and all things that are upon the face of it, yea, and its motion, yea, and also all the planets which move in their regular form do witness that there is a Supreme Creator (Alma 30:44).

Indeed, although there are awesome signs to come, the world has had signs enough. God is witnessed on every hand. All creation testifies of him whose works are everlastingly purposeful works—the Most High God who brought all things—temporal and eternal, tangible and intangible—into existence.

Notes

1 *A New Witness for the Articles of Faith*, 483.
2 Because Mormonism views reality as being on a *material* continuum, the term, "supernatural," is seldom if ever used in theological discussion.
3 As before noted, Mormonism affirms the existence of virtually countless Gods. Collectively, they may be considered essentially omnipotent and omniscient relative to all things—one God in principle. Individually, by association, they possess these attributes relative to their respective kingdoms.
4 *DS*, 1:10; emphasis in original.
5 If law did not operate throughout the universe, it could not properly be referred to as a cosmos, and science as we know it would be an impossibility.
6 See JD, 18:317; 21:321.
7 There is but one divine Spirit. It is variously referred to as the Spirit of God, Spirit of the Lord, Spirit of Christ, Spirit of truth, light of Christ, light of truth, the Holy Spirit, etc. Jesus Christ is the visible embodiment of this Spirit.
8 Henry M. Morris, *The Twilight of Evolution*, 14-15; emphasis added.
9 A major hurdle is the almost sacrosanct Second Law of Thermodynamics. Then too, mathematical calculations, by evolutionists, such as Pierre Lecomte du Nouy, have shown that the time needed for beneficial mutations—the very heart of neo-Darwinism—to produce a human being from even a *living* bacterium exceeds by billions of years the assumed 4.5 billion years presumed for the earth's overall age.
10 Dean L. Overman, 40.
11 Nibley, 23-24; emphasis added.
12 *Improvement Era,* 13:76-81 (Nov., 1909); emphasis added. A portion of this declaration was reprinted in the "*Era*" in Sept. 1925 under the direction of the succeeding First Presidency of Heber J. Grant.
13 Uniformitarianism, first advanced in the 18th century in opposition to catastrophism, received its real impetus in the 19th century from Charles Lyell (1797-1875) in his book, *Principles of Geology.* It is the *a priori* assumption that basic physical processes have remained the same or uniform throughout time. Thus, all phenomena in the astronomic, geologic, and biologic orders are always caused by the same ongoing processes. The recent emergence of plate techtonics is impacting uniformitarianism among geologists.
14 The essence of the cosmological argument is that every *effect* has a prior *cause* and that the *first* of all causes must be God. Philosophical

arguments are not *proofs;* they are arguments, reasonings. Joseph Smith was given *proof* of God's existence and individuality in the First Vision.
15 The teleological argument is that all things give evidence of purpose or design aimed at a meaningful destiny. Hence, if there is design there must be a Designer.
16 See D&C 93:8-10.
17 See 2 Ne.11:4; Hel. 8:24.

CHAPTER THREE

THE POWER OF GOD

A God of Miracles

"God has not ceased to be a God of miracles. . . .if so he would cease to be God" (Morm. 9:15, 19). Miracles are co-eternal with the Almighty because they are the on-going expression of his very nature and powers. They are not extraordinary now-and-then things, they are constant evidences of his benevolent overruling presence. In revealing the "unknown God," Paul testified, "in him we live, and move, and have our being" (Acts 17:28). We cannot "live" in a God of miracles without experiencing them, even though we may do so unwittingly.

The great scientific minds have been among the first to acknowledge the wonders of creation. They realize that the various scientific disciplines are primarily descriptive—knowing much about the *whatness,* but considerably less about the *howness* of it all. It is one thing to map the brain's complexities—identifying those centers controlling bodily actions, emotions and thought processes—and quite another to understand how the brain carries them out. We understand the operations of gravity on an apple, a galaxy, or light, but gravity itself remains something of a mystery. We know what grass is, but we are yet to make a single blade. We catalogue, but we cannot create. So miracles abound because in this, our second estate, we only "know in part." As long as we are ignorant of the *how* behind all things, the universe will remain a miracle to us. And when we learn that *how,* miracles will cease. For a miracle, by definition, is inexplicable; if we know how miracles are produced, they cease to be miraculous. Therefore, for the God of miracles, there are no miracles.

Miracles pose a problem for those who demand a rational, demonstrable explanation for everything. Not understanding how Jesus was able to calm a raging storm, walk upon the sea, or raise Lazarus from the dead, they deny that he did so—"They are just the myths that typically arise around the founders of religions." Why so? Because such things violate known physical laws, and they are not verifiable—meaning, we cannot calm raging storms, walk on water, or raise the dead. Nevertheless, the Lord was, is, and always will be a worker of miracles in the eyes of those whose ways are not his ways and whose thoughts are not his thoughts.

The Power of Man

The achievements of modern science and technology have not come about by accident, nor because human intelligence has experienced a sudden quantum leap. It is because the Spirit of the Lord is deliberately moving upon the minds of men in precise and powerful ways for precise and powerful purposes pertaining to things temporal as well as things spiritual. Brigham Young understood that God is back of it all:

> Instead of considering that there is nothing known and understood, only as we know and understand things naturally, I take the other side of the question, and believe positively that there is nothing known except by the revelation of the Lord Jesus Christ, whether in theology, science, or art. The world receive information and light on great principles of science and knowledge in the arts, *to subserve the hidden purposes of the Almighty*, but they are ignorant of the source from whence it comes to them (JD, 12:207).

Those "hidden purposes" pertain to the salvation of Earth and mankind:

> Every good and perfect gift cometh from God. Every discovery in science and art, that is really true and useful to mankind, has been given by direct revelation from God, though but few acknowledge it. It has been given with a view to prepare the way for the ultimate triumph of truth, and the redemption of the earth from the power of sin and Satan (JD, 9:369).

On the first day of the twentieth century, Joseph F. Smith stated:

> Every unfoldment of the nineteenth century in science, in art, in mechanism, in music, in literature, in poetic fancy, in philosophical thought, was prompted by His Spirit which before long will be poured out upon all flesh that will receive it.[1]

Having learned, as Orson Pratt pointed out, "some of the more gross principles and laws of this fallen creation," man has changed the face of the planet and the lives of us all with those technological advances which bless and curse us today. Impressive as they may be, contrasted with the ways of the Almighty, the achievements of fallen man are still relatively primitive. As in so many things, force is the chief principle man employs in dominating Earth. It does not respond to man's intelligence; it is coerced by it. This has become even more the case as modern technology has enabled man to modify his physical environment in dramatic ways. Machines and explosives level its hills, gouge its mountains, denude its forests, alter the course of its rivers, plunder its oceans, and bury the land under ribbons and splotches of asphalt and concrete.

The issue is not man's undeniable right to use, but his thoughtless abuse, of the earth. Prior to his fall, the immortal

Adam was made the lord of creation. He was given dominion over all things and told to "subdue" the "very good" Earth God had organized. As steward over that creation, Adam was to exercise dominion in righteousness. For Earth is a stewardship, and for every stewardship there must be an accounting. Mankind's indifferent stewardship over this planet, together with his own defiling sins, serve to explain why Mother Earth has long since ceased to yield of her strength, and why she has turned, and will increasingly turn, against man in violent ways. Having sown the wind of callous disregard for God's commandments and the earth's ecological well-being, mankind now faces the prophesied onslaught of latter-day judgments, of nature in rebellion against her despoilers.

The Divine Mind

Whereas "civilized" man tends to view nature as an antagonist to be conquered and exploited, the God of nature is its chief protagonist, willing, not coercing, its obedience. His is the rule of Intelligence or Mind, consequently, the expression "mind over matter" has merit; in the last analysis, the mind does control matter.[2] The eternities are filled with an infinitude of matter, stamped in greater or lesser degree with that omnipresent hallmark of divinity called the Spirit of the Lord. To the degree that this Spirit is present in any particle of matter, it may be said that mind or intelligence is also present. Brigham Young explained:

> There is an eternity of matter, and it is all acted upon and filled with a portion of divinity. . . .[it] is capacitated to receive intelligence. . . .[it] can be organized and brought forth into intelligence, and to possess more intelligence, and to continue to increase in that intelligence (JD, 7:2-3; see 7:285).

Sixteen years later he added: "As far as we are concerned we suggest the idea that there is an eternity of life, an eternity of organization, and an eternity of intelligence from the highest to the lowest grade, every creature in its order from the Gods to the animalcule."[3] The level of agency granted the intelligences on this continuum is directly proportionate to their degree of intelligence—the greater the intelligence, the greater its agency.

The obedience of God's lesser creations to his will results from their being infused with his Spirit to the point that they become, as it were, extensions of it.[4] As we have seen, the Spirit of the Lord permeates all heavenly bodies and is "the power thereof by which they were made" and "the law by which all things are *governed*" (D&C 88:13). Joseph Smith spoke of "the great God who holds this world in its orbit, and who upholds all worlds and all things by his power" (TPJS, 345). It follows, therefore, that in the absence of the unifying influence of the Spirit there could be no cosmos—no order. All would be chaos. Orson Pratt imagined such a state:

> All these universal laws that appear so prominently before us from day to day are nothing more than the operations of that all-wise Spirit which we are told is 'round about and in all things,' and which acts according to certain laws prescribed by the Almighty. . . .Take away this Spirit, and you would immediately see some things going up, others down; some moving horizontally; one portion of the earth would divide from the other; one part would be flying here and another there (JD, 2:340; see 18:317).

In a similar vein, Joseph F. Smith said, "It is by the power of God that all things are made that have been made. It is by the power of Christ that all things are governed and kept in place that are governed and kept in place in the universe" (*GD*, 66). In other words, the omnipresent Spirit of God is the ultimate source of gravitational force. If that Spirit were to be suddenly withdrawn, gravity would cease to operate and every planet, star, and galaxy would disintergrate into chaotic matter. This is only to say that the foundation of the temporal order is literally the spiritual order.

Since the Spirit of the Lord infuses all things, God in principle, not person, is omnipresent in all things.[5] Commenting on God's omnipresence, Brigham Young noted, "there is no portion of space where he is not. There is no element in existence that does not contain him; no matter whether it be in its primitive, or in an organized state, he is through it and round about it. God fills immensity."[6] David asked:

> Whither shall I go from thy Spirit? or whither shall I flee from thy presence? If I ascend up into heaven, thou art there: if I make my my bed in hell, behold, thou art there. If I take the wings of the morning, and dwell in the uttermost parts of the sea; Even there shall thy hand lead me, and thy right hand shall hold me (Ps. 139:7-10).

The Most High possesses holy omnipotence. All lesser orders and gradations of intelligence respond to his will, not by coercion, but by *attraction*. Those men and women who, through their faith and obedience, obtain a celestial exaltation will rule as he rules. To them is the promise: "The Holy Ghost shall be thy constant companion, and they scepter an unchanging scepter of righteousness and truth; and thy dominion shall be an everlasting dominion, and *without compulsory means* it shall flow unto thee forever and ever" (D&C 121:46).

Divine Faith

It may appear erroneous to associate the principle of faith with God who is, to all intents, both omniscient and omnipotent, but faith is essential to any being exercising any degree of agency, including the Almighty himself. He is the embodiment of faith just as he is the embodiment of justice, mercy, love, truth, virtue, knowledge, wisdom and all other eternal principles flowing forth from the Fountainhead of intelligence. So while it is generally assumed that faith ends where knowledge begins, in actuality faith and knowledge are coexisting principles continually interacting upon one another.[7] Knowledge produces faith, and faith in turn produces greater knowledge which then becomes the basis for greater faith, and so on and on to higher and higher levels of knowledge and faith.

One's faith is basically proportionate to the knowledge—valid or not—which produced it and which is, in turn, produced by it.[8] Our limited knowledge gives us the faith to do some things; God's limitless knowledge gives him the faith to do *all* things. He stands at the apex of the faith-knowledge process; his knowledge and faith are perfect and co-exist in him as *indivisible* principles.[9]

In *Lectures on Faith,* faith is defined as that inherent principle of power by which God organizes all things. Indeed, without faith— "the moving cause of *all* action" —God would not be God: "Take this principle or attribute—for it is an attribute—from the Deity, and he would *cease to exist.*"[10]

The principle of faith is whole and holy (being unalloyed with any imperfection) in God. Consequently, when he exercises his faith and releases the infinite power of his mind, all nature obeys. Faith, as it exists in God, is the *alpha* and *omega* of all things. Hence, we read:

> The whole visible creation, as it now exists, is the effect of faith. It was faith by which it was framed, and it is by the power of faith that it continues in its organized form, and by which the planets move round their orbits and sparkle forth their glory. So, then, faith is truly the first principle in the science of THEOLOGY.[11]

The child of faith is works. As a woman may live without children, so may faith exist without works. But the woman is barren, and faith is dormant, if not "dead." Faith lives only when works follow. A living faith alone will produce those works which will endure beyond mortal time. Such is God's faith and, therefore, such are God's works. Only as we partake of his Spirit and manifests a degree of like faith, will our works be infused with immortality.

The Spoken Word of Faith

The most powerful manifestation of faith is found in the spoken word. And the spoken word of God, being the word of faith, is more than verbal communication; it is both the creator and transmitter of power. In explaining what it means to work by faith in God, the seventh lecture states:

> We answer—we understand that when a man works by faith he works by mental exertion instead of physical force. It is by words, instead of exerting his physical powers, with which every being works when he works by faith. God said, 'Let there be light, and there was light.' Joshua spake, and the great lights which God had created stood still. . . . Faith, then, works by words; and with these its mightiest works have been, and will be, performed.[12]

We have little conception of the beauty and majesty of man's primeval language. Far from being a savage tongue consisting of animal-like grunts, the Adamic language of Earth's first inhabitants possessed a power, clarity and comprehensiveness of expression far beyond that of any contemporary tongue. It was for Adam and Eve and their righteous posterity "pure and undefiled" (see Moses 6:5-6, 46). Thus, it was charged with the Holy Spirit, making it powerful and edifying.[13] Unlike modern languages, it was uncorrupted by the presence of alien words or derivations; the curse of Babel was yet to occur. Nor was it defiled by any profanities, obscenities or lascivious expressions.

The Adamic tongue was meat and drink to the human spirit—nourishing, sustaining, and strengthening the whole soul.[14] However, with the passage of time, this food of the spirit, like much food of the body, became ever more corrupt, denatured, and shorn of its vitality and power. Today, language is frequently used to obscure rather than to communicate honest thoughts and feelings, to shock rather than to edify. It is employed as a weapon against the truth calculated to guilefully mislead and, thereby, subvert others—as Satan subverted Eve—to the deceiver's will. On every turn we find glaring examples of this abuse of the divine principle of communication, an abuse compounded by modern technology.[15]

The unadulterated word of God is letter and spirit in union—pure intelligence. When spoken, it can be so "powerful and great" that it cannot be expressed in written form (see Ether 12:25). And when written under the inspiration of the Almighty, it can be so infused with his Spirit that—as was the case with the revelations received by the brother of Jared—they cause "the overpowering of man to read them."[16] An assembled multitude of Nephites and Lamanites heard even more sublime language as it fell from the lips of the purest of men, the resurrected Son of God:

> The eye hath never seen, neither hath the ear heard, before, so great and marvelous things as we saw and heard Jesus speak unto the Father. And no tongue can speak, neither can there be written by any man, neither can the hearts of men conceive so great and marvelous things as we both saw and heard Jesus speak (3 Ne. 17:16-17).

The pre-mortal Christ told the brother of Jared: "And at *my command* the heavens are opened and are shut; and at *my* word the earth shall shake; and at my command the inhabitants thereof shall pass away, even so as by fire" (Ether 4:9). The Only Begotten was the earthly embodiment of the mind, the faith, and the power of the Father— "the Word" which was "made flesh" (John 1:1, 14).[17] He is the instrument by which the Father translates his will into tangible action—the "word of my power" (Moses 1:32, 35; 2:5), which is "the power of my Spirit" (D&C 29:29-30). "Word" and "Spirit" are, therefore, virtual synonyms.[18] As we have seen, the Father's words and works are endless, and they are all fulfilled in and through the Son who said: "I am the same which spake, and the world was made, and all things came by me" (D&C 38:3).[19] Jehovah told Abraham:

> I stretch my hand over the sea, and it obeys *my voice;* I cause the wind and the fire to be my chariot; I *say* to the mountains—Depart hence—and behold, they are taken away by a whirlwind, in an instant, suddenly (Abr. 2:7).

The very particles of Earth are in harmony with, and therefore, subject to his will. He speaks and the earth obeys. "By the power of his voice," mountains become valleys:

> Yea, by the power of his voice doth the whole earth shake. . . .Yea, and if he say unto the earth—Move—it is moved. . .Yea, if he say unto the earth—Thou shalt go back that it lengthen out the day for many hours—it is done. . . .if he say unto the waters of the great deep—Be thou dried up—it is done (Hel. 12:11-16).[20]

The Rebel

The term "obey" is not one a scientist would use in connection with his manipulations of matter and energy. However, it is used in scripture to express the relationship between God and his creations. But man is a rebel where God is concerned.

Although we humans constitute the highest order of intelligence in existence, as mortals we have the greatest difficulty harmonizing our wills with that of the Supreme Intelligence. Brigham Young observed, "Man is the only object you can find upon the face of the earth that will not abide the law by which he is made" (JD, 9:141). Perhaps the very superiority of man's intelligence combined with the gift of agency and the complexity of human nature are at the root of the problem. Then there is the devil!

Endowed with a high degree of moral agency when possessing knowledge, man is free either to obey or disobey his Creator as he chooses. An all-wise Father wills life-giving principles for his children but, like many mortal parents, he must sadly acknowledge, "Men do not always do my will" (D&C 103:31). Hence, scripture declares mankind to be foolish, vain, proud, devilish, boastful, and rebellious toward God: "For they do set at naught his counsels, and they will not that he should be their guide" (Hel. 12:6). Yet we pit our agency against the agency of God with inevitable pain and sorrow.

Such is not the case with those lesser orders of life and intelligence which lack knowledge and capacity and, therefore, agency in its higher and more extensive forms. They are essentially creatures of instinct, other-directed rather than self-directed.[21] Being so, they are responsive to the divine will and they are secure in the divine order of things.[22] Thus, nature is obedient to nature's God.

But man is a child, not a creature, of that God. As such he is above and apart from nature; he is free as nature can never be. If man would be free, as Christ is free, he must emulate nature's example and humble himself before nature's God. Only in this way can he enter into union with the Father and become a co-ruler of his works. Those who choose to disregard nature's example will be out of harmony with the eternal order of things and, therefore, unfit for rule in that order. They forfeit the kingdom, the power, and the glory—the exaltation and eternal lives—the Father bestows upon his most obedient children.

Men of Power

There are remarkable examples in Holy Writ of the power of the spoken word manifest through mortal men. Enoch, that "slow of speech" patriarch, was promised, "all thy words will I justify; and the mountains shall flee before you, and the rivers shall turn from their course" (Moses 6:34). This divine promise was fulfilled to the letter:

> Enoch spake the word of the Lord and the earth trembled, and the mountains fled, even according to his command; and the rivers of water were turned out of their course... so powerful was the word of Enoch, and so great was the power of the language which God had given him (Moses 7:13; see JST, Gen. 14:26-32).

Moses drew down the power of God in the ten plagues and the dividing of the Red Sea, only to be later chastised for disobedience. When the Israelites clamored for water, Jehovah told Moses: "Speak ye unto the rock before their eyes; and it shall give forth his water" (Num 20:8). But instead of using the pure word of faith, Moses struck (coerced) the rock. As a merciful penalty for his disobedience, Moses was denied the burden of bringing Israel into the land of Canaan.

Joshua's faith produced an unprecedented miracle cited in Lectures on Faith. After praying for help, Joshua was inspired to command: "Sun, stand thou still upon Gibeon; and thou, Moon, in the valley of Ajalon." And the sun and the moon obeyed Joshua so that "there was no day like that before it or after it, that the Lord hearkened unto the voice of man: for the Lord fought for Israel" (Josh. 10:12-14).[23] The Lord responded to Joshua's faith.

Moroni wrote of an earlier time when, in like manner, the Lord hearkened unto the voice of another man of faith: "For the brother of Jared *said* unto the mountain Zerin, Remove—and it was removed. And if he had not had faith it would not have moved; wherefore *thou workest* after men have faith" (Ether 12:30).

Nephi told his unbelieving brothers: "If he should command me that I should *say* unto this [the Arabian Sea], be thou earth, it should be earth; and if I should *say* it, it would be done" (1 Ne. 17:50). Little wonder Nephi wrote that he was not "mighty in writing like unto speaking" (2 Ne. 33:1).

While men often ignore the inspired word, nature does not. Jacob, Nephi's younger brother, wrote that their faith was so great "that we truly can *command* in the name of Jesus and the very trees obey us, or the mountains, or the waves of the sea" (Jacob 4:6). Jacob further tells us that God created the earth and man *"by the power of his word."* He then asks the question:

> Wherefore, if God being able to *speak* and the world was, and to *speak* and man was created, O then, why not able to command the earth, or the workmanship of his hands upon the face of it, according to his will and pleasure (Jacob 4:9).

No one manifest this principle so frequently or more directly than did Jesus. The four gospels record more than forty of his miracles, and undoubtedly there were many others of which we have no record (see John 21:25). He healed the sick, crippled and infirm, restored sight to the blind, hearing to the deaf, life to the dead, cast out devils, turned water to wine, fed the five thousand, and calmed raging storms. In all but two very minor instances, his recorded miracles were accomplished without any physical action on his part.[24] Such is the power of the spoken word when that word is in harmony with, and a manifestation of the mind and will of the great Jehovah.

Divinity is bestowed upon those men and women who are sanctified and exalted in the highest degree of the Celestial kingdom. They have become like unto the Father and the Son and are made perfect in light and truth—divine intelligence. As God is love, so have they become living embodiments of this same defining principle of holiness which, when coupled with a fulness of Priesthood, qualifies them to exercise the power of God the only way it can be excercised—in righteousness.

Notes

1 Clark, 3:335.
2 In Mormon theology, all existing realities possess materiality. Mind, spirit, and intelligence should be thought of as higher expressions of matter rather than as abstract or immaterial qualities (see D&C 131:7). Joseph Smith spoke of element or chaotic matter as that "in which dwells all the glory" (TPJS, 51).
3 General Conference, 8 October 1875; see JD, 3:277.
4 This doctrine should not be confused with animism, the belief that individual, *immaterial* spirits are the life force in both animate and inanimate objects.
5 For a thoughtful treatment of this concept, see B.H. Roberts, "The Nearness of God," a discourse delivered in the Salt Lake Tabernacle, 15 Mar. 1914.
6 Discourse of 12 Feb. 1854, Church Archives. Brigham Young is not referring to God's personal spirit which is an inseparable part of his resurrected nature, but to the omnipresent Spirit of the Lord. See D&C 88:12-13, 41.
7 This does not contradict Alma's statement: "Faith is not to have a perfect knowledge of things." (Alma 32:21.) While knowledge (via experience) of a given thing eliminates the need for faith in *that* thing, it does not eliminate the need for faith pertaining to things which have not yet been experienced or achieved. Jesus was obliged to draw upon his own faith each time he performed a miracle. Thus, we may have a perfect knowledge of that which has been achieved, while at the same time exercising faith in that which is *yet* to be achieved. See Alma 32:34-36.
8 See Rom. 10:14.
9 God's faith is non-contingent, meaning it is not dependent upon anyone or anything other than himself. It is perfect confidence, based upon perfect knowledge, that he can do anything he chooses to do; see Abr. 3:17; D&C 3:1-3.
10 *Lectures on Faith,* no. 1, pars. 12, 16; emphasis added
11 *Lectures on Faith,* no. 7, par. 5; see Heb. 11:3.
12 *Lectures on Faith,* no, 7, par. 3; emphasis added.
13 Nephi wrote, "when a man speaketh by the power of the Holy Ghost the power of the Holy Ghost carrieth it unto the hearts of the children of men" (2 Ne. 33:1).
14 See Deut. 8:3; Amos 8:11; Matt. 4:4.
15 In addition to the printed word, radio, television, motion pictures, the telephone and now, the internet are employed to wage war against com-

mon decency and moral law. Little wonder, Satan is "the prince of the power of the air" (Eph. 2:2).
16 Ether 12:24. The inspiration of the Spirit of the Lord in the writing of scripture is the critical ingredient which distinguishes true from false revelation. This is why William McLellan, who had criticized the revelations of Joseph Smith, could not successfully meet the Lord's challenge to produce one of his own equal to the most inferior of those in the *Book of Commandments.* See D&C 67:4-9; HC, 1:224-26.
17 Jesus Christ is "the Word, even the messenger of salvation" (D&C 93:8).
18 Since the power of the Father is manifest in and through his Son, there is no contradiction in identifying the "word" with both Christ and the Spirit (see D&C 29:30; 93:8-10).
19 See 1 Ne. 17:46; Morm. 9:17.
20 Earth's obedience to the divine will serves to explain why the Lord said, "the earth abideth the law of a celestial kingdom."
21 The animal kingdom is characterized by gradations of intelligence which allow for greater or lesser degrees of instinctual behavior; see Rev. 4:6-9; TPJS, 291-92.
22 Animals fill the measure of their creation and will enjoy "eternal felicity" in the kingdom of God; see D&C 77:3-4.
23 The reality of this miracle is affirmed in Hel. 12:14-15. Either by personal revelation or from reading the Brass Plates, Mormon assumed that the rotation of the earth was reversed! To say the least, this would defy the known laws of physics. The Lord is not bound by those laws, but, as John A. Widtsoe wrote, "Even limited human knowledge suggests several simpler methods–refraction and reflection of light, for instance, by which the extension of daylight might be accomplished. Divine power may stop the rotation of the earth, let that be clearly accepted, but it certainly may have at its command other means for extending the hours of light in a day." (ER, 113). The birth of Christ was accompanied by such a miracle when a night without darkness was experienced in the Western hemisphere so that for a period of hours the *entire* earth was bathed in light! See Hel. 14:3-4; 3 Ne. 1:15, 19.
24 See Mark 7:32-34; John 9:6. In both instances it is obvious that Jesus did what he did with the faith of those present in mind, not because his actions were actually necessary.

Part Two

The Earth

Chapter Four

Why The Earth?

Earth's Measure of Creation

God organized Earth with his children in mind—it exists because we exist. Like us, it is eternal in its design. The Lord declared, "the earth abideth the law of a celestial kingdom, for it filleth the measure of its creation, and transgresseth not the law" (D&C 88:25). That is, Earth is fulfilling its divine purposes and will become what it was created to become. Those purposes are centered in the Patriarchal Order of the Father—the *Man* of Holiness. *Man*kind is the Godkind, the highest order of intelligence in existence —there are no exotic creatures lurking somewhere in the universe equal to, much less superior to, Mankind.

Moses, transfigured in glory, stood in the presence of God, and talked with him "face to face" (Moses 1:31). Perplexed by the existence of the many other inhabited worlds he had seen following his humbling vision of this one planet, Moses asked, "Tell me, I pray thee, why these things are so, and by what thou madest them?" The Son, speaking for the Father, replied:

> . . .worlds without number have I created; and I also created them for mine own purpose. . . .And as one earth shall pass away, and the heavens thereof even so shall another come; and there is no end to my works, neither to my words (Moses 1:30-33).

Earth was far from unique; it had countless companions; and it was destined to "pass away" to a higher state of glory.

But why this endless procession of worlds; what possible benefit were they to the Almighty; were they not sheer excess? His answer summed up the essence of his own being in a single sentence: "For behold, this is my work and my glory—to bring to pass the immortality and eternal life of man" (Moses 1:39). Herein is revealed the highest expression of the law of consecration, the profound love and selflessness that is at the core of the Father's being. These worlds were not for himself, but for his children. Possessing all, the Father is continually seeking to share the riches of eternity with intelligences yet unorganized, spirits yet unborn—bringing them into organized existence, endowing them with his own attributes and powers, and providing the way whereby they might attain unto a fullness of his glory and his joy! God is love and love is the reason the family of the Supreme Patriarch is endlessly expanding throughout the eternities. It is the motivation for the on-going organization of new worlds. Each spirit generation must claim physical bodies and experience its own earthly probation in pursuit of that for which Jesus Christ came, and for which he was sacrificed: the more abundant life of truth, virtue, dominion, and happiness (see John 10:10). Thus, the Father's course is "one eternal round."[1]

The infinite in God and his works is a presently incomprehensible article of faith. He declared, "my works have no end, neither beginning" (D&C 29:33).[2] Consequently, there must be an infinite amount of matter and space. Were God's labors limited, his works, however vast, would be finite in number and his immutable promise of "a continuation of the seeds [posterity] forever and ever" could not be fulfilled.[3] This planet is but one tiny link in an unbroken chain of inhabited worlds reaching back far beyond Kolob and extending on into endless time.

In Their Steps

Mortality has a number of purposes, but they are all centered in one universal objective: eternal life. At some point in our first estate we became aware of the perfection, powers, and privileges of the Gods—men and women who had achieved exaltation. Our own divine parents had left that estate for an earlier earth where they acquired physical bodies and, overcoming all, proved worthy of a like exaltation.[4] They invited us to follow in their steps. Although not all of their children would achieve a like glory, still, there would be glory for all. All would appreciably benefit from entering mortality, the second estate, however brief or trying their stay. To remain forever in the first estate as unembodied spirits was neither possible nor desirable. Even heaven, ere long, would become a dead-end street. So we willingly came to this world as our Parents, eons ago, had journeyed to another, more ancient, one.

And we came for the same basic reason: to work out our salvation, as our Father worked out his— "with fear and trembling." (TPJS, 347; Philip. 2:12). But we did not buy "a pig in a poke;" we had a measure of understanding of what mortality would entail. The Lord foreknows the course of every life and foreordains the course of some.[5] Indeed, that the course of our lives was foreknown not only to God, but to some extent to ourselves was the conviction of President Joseph F. Smith:

> He [Jesus Christ] possessed a foreknowledge of all the vicissitudes through which he would have to pass in the mortal tabernacle. . . . If Christ knew beforehand, so did we. But in coming here, we forgot all, that our agency might be free indeed, to choose good or evil, that we might merit the reward of our own choice and conduct (GD, 13-14).[6]

And we came, not because we were forced to do so, but of our own free will, longing to be added upon. It was the ultimate adventure!

A Second Birth

Eastern religions, such as Hinduism and Jainism, espouse reincarnation and transmigration of souls,[7] the objective being eventual liberation from physical realities by escaping into Nirvana. This is the opposite of the truth: the physical body is the very key to salvation and self-fulfillment. Judaism and the traditional Christian faiths, Catholic and Protestant, have no doctrine of a pre-mortal existence for mankind. They believe that the first human spirit or soul originated with Adam when "the Lord God formed man of the dust of the ground, and breathed into his nostrils the breath of life; and man became a living soul" (Gen. 2:7). Thus the doctrine that mortality is actually a second birth for spirits procreated by resurrected parents in a former (first) estate is quite unique to Mormonism.

Entering the "second estate" meant gaining a tabernacle of flesh and, therefore, realizing fuller self-hood. The Father, a resurrected being of spirit and element, is a microcosm of all materiality. There is no principle of life existing anywhere in the eternities that is not represented in him. Having subdued and sanctified all materiality, his perfection is absolute and his dominion limitless —he has a celestial "fullness of joy."[8] His sons and daughters aspired to become like him and their divine Mother. But they understood that this sublime goal was possible only when the spirit was clothed upon with a physical tabernacle. There is no fullness of joy, for an unembodied spirit.[9]

> For man is spirit. The elements are eternal, and spirit and element, inseparably connected, receive a fullness of joy; And when separated, man cannot receive a fullness of joy (D&C 93:33).

As spirits, we came to earth to be "added upon" with element, becoming living souls. All this, in pursuit of happiness—the "object and design of our existence" (TPJS, 255). To the extent that we subject element, the flesh, to the will of our spirit, and our spirit to the will of God, we find our own respective "fullness of joy."

True evolution had its beginnings in the unorganized primeval spirit state. From that state all things unfold, not randomly or capriciously, but in accordance with their respective measure of creation: the God-determined purpose or end for which they were organized as spirit entities. Even as the physical world is a reflection of its spirit counterpart, so is our physical body patterned after the basic likeness of our spirit. In appearing to the brother of Jared, Jesus Christ asked him:

> Seest thou that ye are created after mine own image? Yea, even all men were created in the beginning after mine own image. Behold, this body which ye now behold, is the body of my spirit; and man have I created after the body of my spirit (Ether 3:15-16).

We came forth male or female at spirit birth and remain male or female forever thereafter. Those problems associated with a confused sexual identity will, in due time, be corrected along with all other physical, psychological and moral imperfections.

A Living Earth

Earth is altogether unique among the planets of this solar system. It alone has the delicately-balanced environment suitable for life as we know it. It alone possesses all of the essential elements necessary for creating, developing, and sustaining physical bodies capable of a glorious quickening unto immortality.[10] These elements, wonderfully combined, constitute the third essential component of our eternal being—the component which transforms our unembodied spirit into a living soul.[11] This component is "formed from the dust of the ground"—the good earth. God told Adam,

> By the sweat of thy face shalt thou eat bread, until thou shalt return unto the ground—for thou shalt surely die—for out of it wast thou taken: for dust thou wast, and unto dust shalt thou return (Moses 4:25; see Gen. 3:19).

Earth is the mother of mothers; every living organism is comprised of elements taken from her body. Whether man, the denizens of land, sea, and sky or a microscopic organism—all are extensions of a living earth, a living soul, a life-producing planet. Orson Pratt asked the rhetorical question, "What! is the earth alive too?" He responded, "If it were not, how could the words of our text [Isaiah 51:6] be fulfilled, where it speaks of the earth's dying? How can that die that has no life?" (JD, 1:281) In a similar vein, Heber C. Kimball argued, "if a woman will not produce when she is dead, then the earth cannot produce living things if it was dead. . . .The earth is alive. If it was not, it could not produce" (JD, 6:36) John A. Widtsoe wrote, "Certainly, the earth on which we live is an imperishable, living organism" (ER, 148). Thus, the principle of biogenesis (life begets life) also applies to Mother Earth. Because she lives, we live. And she will continue to provide the critical elements from which the physical bodies of mankind and the "generations of the heavens and the earth" are fashioned until the end of time.[12]

A pre-determined number of male and female spirits are destined to enter mortality on this planet. Marriage was ordained of God to provide these spirits with a righteous and secure entry into their second estate. This, that "the earth might answer the end [purpose] of its creation; And that it might be filled with the measure [number] of man according to his creation before the world was made" (D&C 49:16-17). Not until the last spirit destined for this planet has claimed a physical body will the earth "answer the end of its creation." Nothing men, women, devils, doctors, or politicians do will prevent it. No spirit will be denied a physical body because of the actions of others; there will always be willing and waiting parents.

"We Will Prove Them Herewith"

Abraham's record refers to "the intelligences that were organized [the spirits that were begotten] before the world was." Within this multitude of males and females numbering in the tens of billions were "many" who were "good" or God-like—the "noble and great." "These," said God, "I will make my rulers."[13] But the eternal destiny of every child of God was dependent upon the manner in which all available commandments were honored, and the response of each, whether in life or death, to the gospel of Jesus Christ.

> We will prove them herewith [earthlife], to see if they will do all things whatsoever the Lord their God shall command them. And they who keep their first estate shall be added upon; and they who keep not their first estate shall not have glory in the same kingdom with those who keep their first estate; and they who keep their second estate shall have glory added upon their heads for ever and ever (Abr. 3:25-26).[14]

Would absence make the hearts of the Father's children grow fonder? Or would it be a case of "out of sight, out of mind"? Would they keep his commandments as faithfully on Earth as they had done in heaven? Would they be as successful in rejecting the overtures and blandishments of Lucifer in a new world as they had been in the old one? Would they continue to grow from grace to grace in light and truth in their quest for eternal life? Would they learn to rule gross matter and not be ruled by it? Would they, in acquiring a tabernacle of element, the flesh with its appetites and passions, be able to subdue it to the rule of the spirit? Would they, with divine help, sanctify themselves so that they could enjoy the association of holy beings?[15] The answers to these questions would vary.

How each spirit exercised his or her gift of agency made for great variety in the spiritual, moral, and intellectual composition of the heavenly host. Consequently, before Earth was ever organized, the human family had already become highly diversified insofar as intelligence, worthiness, talents, and achievements were concerned. "The spirits in the eternal world," said Joseph Smith, "are like the spirits in this world" (TPJS, 305; see 297). Melvin J. Ballard, reflecting this doctrine, said:

> I would like you to understand that long before we were born into this earth we were tested and tried in our pre-existence. . . .There are no infant spirits born. They had a being ages before they came into this life. They appear as infants, but they were tested, proven souls.[16]

Joseph Fielding Smith maintained, "The character of our lives in the spirit world has much to do with our disposition, desires and mentality here in mortal life" (DS, 1:160). Thus, the diligence with which we learned the lessons of the first estate in large measure determined our fundamental character and personality and, therefore, our opportunities for spiritual and intellectual growth, progress and service in this second estate.[17] The operative law is:

> Whatever principle of intelligence we attain unto in this life, it will rise with us in the resurrection. And if a person gains more knowledge and intelligence in this life through obedience than another, he will have so much the advantage in the world to come (D&C 130:18-19).

This law also applies to our former existence; when we came to earth, we brought with us whatever principles of intelligence, whatever divine qualities, we acquired in our first estate. The spiritual character of most of mankind are largely a reflection of this fact. In the main, this is all to the good. Brigham Young counseled, "Never try to destroy a man. It is our mission to save the people, not to destroy them. The least, the most inferior spirit now upon the earth, in our capacity, is worth worlds" (JD, 9:124). The writer heard Elder Matthew Cowley express a similar view: "There isn't a man living who isn't greater than his sins." This explains why virtually all mankind will be saved; they are greater than their sins. The fundamental goodness and worth of many souls may lie hidden beneath an overburden of sin, ignorance, and circumstance, but godly qualities and genuine worth are there nonetheless.

Spiritual "scholarships" were given to the Father's most meritorious sons and daughters—the noble and great ones—whom he determined to make his "rulers" on earth both in time and eternity.[18] God's great servants from Adam to the present day were foreordained in the councils of heaven. Speaking of his own prophetic calling, Joseph Smith declared:

> Every man who has a calling to minister to the inhabitants of the world was ordained to that very purpose in the Grand Council of heaven before this world was. I suppose that I was ordained to this very office in that Grand Council (TPJS, 365).

In his vision of the redemption of the dead, President Joseph F. Smith saw Joseph Smith and a number of other leaders of the Church in the spirit world: "I observed that they were also among the noble and great ones who were chosen in the beginning to be rulers in the Church of God" (D&C 138:55). If those who were foreordained to be the elect of God keep their second estate with the same diligence manifested in their first estate, they will have proven themselves in "all things whatsoever the Lord their God shall command them" (Abr. 3:25).

Since certain spirits were judged "noble and great," it follows that others were judged less so, being found wanting in spirituality and obedience.[19] The gross infidelity of some toward God in this world may very likely reflect an earlier spirit of rebelliousness. Lucifer is the classic example. His rebellion being total, his judgment was final—he and his followers were denied the key to eternal life, a physical tabernacle. Consequently, they did not obtain a "glory" in any earthly "kingdom."[20] Others, while not joining Lucifer's rebellion were, for whatever reasons, not as successful in keeping (or magnifying) their first estate as were others. As a group, these less-achieving spirits "shall not have glory in the same kingdom with those who keep their first estate."[21]

The foregoing must be qualified. Through the ages many spirits have lived on earth under conditions which precluded their ever hearing of the true God or of the gospel of Jesus Christ. Joseph Smith referred to such individuals as "heathens."[22] Being without divine law, they will be judged without such law in accordance with the moral and spiritual understanding they possessed and the lives they led.[23] However, every spirit was taught the gospel of Jesus Christ in the first estate and is taught it again either in mortality or in the spirit world following death. Thus, every child of God is provided with a full opportunity at some point prior to the last judgment to qualify for the highest degree of glory of which he or she is worthy and capable. Where the Lord is concerned, there are no "lost sheep."

Earthly Opposition

In time we shall discover that the greatest joy may grow out of the deepest sorrow. Mortality introduced mankind to a new dimension of the eternal principle of opposition in all things—the opposition between Heaven and Earth, light and darkness, spirit and flesh. It was not until we were subjected to an alien condition of privation, sickness, and death that we could begin to rightly judge the merits of our former estate. The Garden of Eden was loveliest when Adam and Eve looked back upon it from a fallen world. Heaven will be all the more heavenly now that it has been surrendered for a season on Earth. Immortality will be all the more a wonder for our descent into death. Wilford Woodruff observed:

> It has no doubt been a marvel many times in the minds of men and women, why God ever placed men and women in such a world as this, why he causes his children to pass through sorrow and affliction here in the body. . . .[the Father] has placed us here that we may pass through a state of probation and experience, the same as he himself did in his day of mortality. . . . If we never taste the bitter, how will we know how to comprehend the sweet? If we never partake of pain how can we prize ease? And if we never pass through affliction, how can we comprehend glory, exaltation and eternal blessings? (JD, 18:32-33).

Spiritual growth and development is the underlying reason we are obliged to live under conditions of injustice, suffering, privation, and strife on this earth—conditions that are the antithesis of all we had previously known. Godliness is forged in the furnace of affliction.

Therefore, another objective of mortality which goes beyond a deeper appreciation of the goodness of God is the quest for a fullness of glory. In mortality our spirits are veiled in flesh. This subjugation of the spirit was necessary if we are to have those unique experiences which mortality alone can provide. As Orson Pratt noted: "We learn by our experience many lessons we never could have learned except we were tabernacled in the flesh" (JD, 14:242).[24] Such is the paradox: we need the body to progress but, in its fallen condition, it is a hindrance to the spirit's efforts to do so. "We have to fight continually, as it were," said Brigham Young, "to make the spirit master of the tabernacle, or the flesh subject to the law of the spirit" (JD, 9:287).

The only way the Father's children could acquire his understanding was for them to undergo such experiences as would give them the peculiar knowledge of good and evil, of the opposition in all things, he possessed. He is able to comprehend the trials and sufferings of his children because he had similar experiences in an earlier stage of his own existence. Brigham Young maintained that it would be foolish to petition a God "who never had any experience, in the adverse fortunes of mortal life." The Lord can respond compassionately to our pleas for support in times of tribulation because he has suffered like trials:

> If he has received his exaltation without being hungry, cold, and naked; without passing through sickness, pestilence and distress, by the same rule we may expect to be exalted to the same crown, to the same glory and exaltation. . . . No man who is well instructed in the things of God would exclaim; 'The God I serve has received his exaltation without suffering, and learning by his faithfulness.' If God has received his exaltation in this way, He cannot in right, and in justice, call upon us to earn it by suffering, for He would then require that of us He Himself was not subject to.[25]

The Savior had to drain the bitter cup his Father gave him because there was no other way for him to fulfill the plan of salvation. The bitter-sweet cup of mortality appears to be likewise necessary for those who aspire to become joint-heirs with Jesus Christ in all that the Father possesses. The Prophet Joseph declared, "Men have to suffer that they may come upon Mount Zion and be exalted above the heavens" (TPJS, 323).

In coming to Earth, the Son of Man descended into the deepest hell that he might rise to the highest heaven. In doing so, "he comprehended all things, that he might be in and through all things, the light of truth" (D&C 88:6). Recounting the many trials Joseph Smith had endured, the Savior assured him, "all these things shall give thee experience, and shall be for thy good. The Son of Man hath descended below them all, Art thou greater than he?" (D&C 122:6-8). Although we are called upon to take up our cross,[26] whatever it might be, it is not, and can never begin to be, the burden borne by the Savior. The cross fits the man or the woman, be they greater or lesser. The Lord does not impose any trial upon us greater than our capacity to endure (see 1 Cor. 10:13). However, the devil is not so obliging; he will bring us to destruction if we surrender our agency to him.

Earth life is exaltation's obstacle course; it provides those hard experiences which alone develop those qualities and capacities essential for those who would follow in the steps of the Gods. In striving to do so, we must not expect the way to be easy, we must not expect constant spiritual support. Spiritual trials come in the absence, not the presence, of the Lord. Orson Pratt cautioned against unrealistic spiritual expectations:

> But all these great Prophets, Seers and Revelators had to experience their seasons of darkness and trial, and had to show their integrity before God in the midst of the difficulties they had to encounter. Shall the Latter-day Saints despond, then, because they may have seasons of darkness, and may be brought into trials and difficulties? No! Let us be steadfast, holding fast to the rod of iron—the word of God—and to our honesty, integrity and uprightness, that God may be well pleased with us whether we have much or little of the Spirit. I do not know how we could have many trials, if we were all the time filled with the Spirit and continually having revelations from on high (JD, 15:236-37).

The Worthy Dead

There is a class of spirits consisting of those who would have received the gospel in the flesh had the opportunity been afforded them. It is this class of spirits for whom the doctrine of salvation for the dead is primarily designed. The Lord informed Joseph Smith:

> All who have died without a knowledge of this gospel, who would have received it if they had been permitted to tarry [until the gospel was available], shall be heirs of the celestial kingdom of God; also all that shall die henceforth without a knowledge of it, who would have received it with all their hearts, shall be heirs of that kingdom; for I, the Lord, will judge all men according to their works, according to the desire of their hearts (D&C 137:7-9; see 128:5).

It is obvious that the hearts of all men and women are known to the Lord well before they enter mortality. From his standpoint, the book of life was written before it was lived. We come to Earth to "prove" ourselves to ourselves so that in the day of judgment we might acknowledge "all of his judgments are just" (Alma 12:15).

This revelation also makes it clear that Earth, including its spirit world, is not only a proving ground for the living but for the dead as well. Our preparatory or probationary state begins with birth and ends with "the night of darkness wherein there can be no labor performed."[27] The afterlife (a continuation of the second estate), is therefore, a segment of that period in which the message of salvation is proclaimed to those spirits held captive behind the gates of hell in the prison-house of death.[28] Since the final judgment is based upon the deeds done in the body, no losses can be incurred after death. The dead will, by repenting and acknowledging Jesus Christ as their Lord and Savior, qualify for their merited salvation;[29] they cannot sin after dying so as to jeopardize it. The level of divine law obeyed in the first and second estates determines the level of divine law (and, therefore, divine blessings) which can be attained after the resurrection (see D&C 88:20-39).

The Destiny of Children

In addition to those who lived and died in ignorance of the plan of salvation, literally billions of the Father's children have passed away near birth, in infancy, or in early childhood having no earthly opportunity to "prove" themselves vis-a-vis the commandments of God.[30] They accomplished the primary purpose of mortality—a physical body—the one universal objective of every earthbound spirit, all else is variable and relative. In being spared the necessity of a mortal probation,[31] they will never be "tried and tested," never be subjected to the temptations of Satan or the trials and vicissitudes associated with this world. Elder Joseph Fielding Smith explained:

> *Satan cannot tempt little children in this life, nor in the spirit world, nor after their resurrection.* Little children who die before reaching the years of accountability will not be tempted; those born during the millennium when Satan is bound and cannot tempt them, "shall grow up without sin unto salvation" (DS, 2:57; emphasis in original).

Young children are "alive in Christ, even from the foundation of the world" (Moro. 8:12). At death, having left the mortal body, they enter the spirit world as *adult* spirits and dwell among the righteous.[32] This merciful dispensation to so vast a number of spirits was a necessary part of the plan of salvation which takes into account the moral agency of mankind and the diverse consequences of it. Whatever sins we committed as spirits in our first estate were remitted through the atonement,[33] consequently we bring no burden of guilt into the second estate: "Every spirit of man was innocent in the beginning; and God having redeemed man from the fall, men became again in their infant state, innocent before God." (D&C 93:38).[34] There is no double jeopardy; we are subject to judgment only for the deeds done in the body during the years of accountability.[35] Such deeds not exisiting, no judgment can be imposed on this class of spirits.

In the previously cited vision of 21 January 1836, Joseph Smith "beheld that all children who die before they arrive at the years of accountability, are saved in the celestial kingdom of heaven" (D&C 137:10). While some have suggested that this applies only to the offspring of the saints or to those "born under the covenant," the Prophet's statement is generally interpreted as an assurance that all such spirits are assured of celestial glory. The declaration appears to be an addendum to the preceding statement that mankind will be judged "according to their works, according to the desire of their hearts." The salvation of young children (who are actually adult spirits) is surely in harmony with this principle. If so, then it would seem that, had they lived, these spirits would have accepted the gospel and would have qualified for salvation "in the celestial kingdom of heaven."

The theological implications of this doctrine are profound since it not only exempts billions of spirits from being subject to the principle expressed in Abraham 3:24-25, but also assures them of salvation in the highest kingdom. If this is correct, it is obvious that the first estate was far more significant and predictive for many men and women than we have assumed. While determining the destiny of one vast portion of the human family, it left unresolved that of another. Why this colossal judgment? We are but babes in the things of God, and this leads us to over simplify his works and ways.

The doctrine saving all deceased young children in the celestial kingdom has been objected to on the grounds that the accident of premature death is hardly a just reason for saving anyone. Why should those spirits who are subjected to the trials of mortality risk all for salvation while other spirits, risking nothing in this estate, are assured of a celestial glory? The question implies that the Lord acted arbitrarily in the matter. But he is no respecter of persons; nor is he capricious—salvation on any level is not a matter of chance. As with every other blessing, it is based upon obedience to irrevocable law. We have been working out our salvation from the very beginning of our association with the Father. His judgements are a response to our moral agency.[36] Indeed, without such agency we could not have chosen to "keep" (honor) or lose the blessings of the first estate.

The first estate was a dynamic, not passive, experience. Aside from all else that transpired there, the Father's entire spirit family was obliged to face the demagogic opposition of Lucifer and thus be proven as to their essential spiritual integrity. So compelling were the issues in the first estate that, for many, a decisive judgment was made then and there, while for many others, for various reasons, judgment was deferred following their works in the second. It is fruitless to conjecture why individual spirits are in one group rather than the other or to argue their relative worthiness. It is certainly incorrect to say that only the good die young. The plan of salvation required many of his noblest sons and daughters to be foreordained to live to maturity to parent their unborn brothers and sisters and to fulfill many other missions related to the salvation of man. Who was more worthy than the Son of God?

The basic question— "Are little children heirs of the celestial kingdom because they die, or do they die because they are heirs of the celestial kingdom?" —brings us back to the issue of God's culpability in their deaths. Does he cause the tragic deaths of a favored class of adult spirits before they attain the age of accountability in the flesh? Assuredly not! A moment's reflection on the historic practices of foeticide and infanticide, together with the many other immoral circumstances under which children die, should make it obvious that God has no part in these crimes against innocence.

However, the Lord foreknows all things, and appoints his children, with their consent, to Earth in those times, places, and circumstances most appropriate for them and in accordance with his wisdom, justice, and mercy. Thus, the Father is obliged to work with man's agency, but he is never a party to its wrongful use. Jesus warned: "Woe unto the world because of offences! for it must needs be that offences come; but woe to that man by whom the offence cometh!" (Matt. 18:7).

At the same time, Joseph Smith suggested that the very temperament of some of the Father's children was such as to make mortality an unnecessarily harsh ordeal for them:

> The Lord takes many away even in infancy, that they may escape the envy of man, and the sorrows and evils of this present world; they were too pure, too lovely, to live on earth; therefore, if rightly considered, instead of mourning we have reason to rejoice as they are delivered from evil, and we shall soon have them again (TPJS, 196-97).[37]

Again, such favored treatment must be based on pre-mortal merit, not arbitrary mercy.

The Prophet comforted bereaved parents in saying, "A question may be asked— 'Will mothers have their children in eternity?' Yes! Yes! Mothers, you shall have your children; for they shall have eternal life, *for their debt is paid*" (HC, 6:316; emphasis added). While not defining the term "eternal life," he seemed, in this instance, to equate it with the phrase "saved in the celestial kingdom."

Although the atonement apparently saves little children in that heaven, of itself, it does not exalt them there.[38] That issue is yet to be resolved by specific revelation from the Lord. Brigham Young remarked that such children "are to receive an exaltation"[39] but then went on to qualify the statement by adding, "Should it be that they are not capable of dwelling as fathers and mothers—of becoming enthroned in glory, might, majesty, and power—they may enjoy to a fullness, and who can enjoy more?"[40] This reflects his view that a "fullness" is determined by one's capacity and that men are equal when their respective capacities are "filled to overflowing."[41] Speaking of deceased children, Joseph F. Smith taught:

> They will inherit their glory and their exaltation, and they will not be deprived of the blessings that belong to them. . . all that could have been obtained and enjoyed by them if they had been permitted to live in the flesh will be provided for them hereafter. They will lose nothing by being taken away from us in this way (*GD*, 453).

Since theological terms are sometimes used rather loosely, we must be cautious in making dogmatic statements based on isolated comments. President Wilford Woodruff was uncertain as to the precise state of deceased children after the resurrection:

> Why our children are taken away from us it is not for me to say. . . .Children are taken away in their infancy, and they go to the spirit world. . . .With regard to the future state of those who die in infancy I do not feel authorized to say much. *There has been a great deal of theory, and many views have been expressed on this subject,* but there are many things connected with it which the Lord has probably never revealed to any of the Prophets or patriarchs who ever appeared on the earth. There are some things which have not been revealed to man, but are held in the bosom of God our Father, and it may be that the condition after death of those who die in infancy is among the things which God has never revealed; but it is sufficient for me to know that our children are saved. . . . .With regard to the growth, glory, or exaltation of children in the life to come, God has not revealed anything on that subject to me, either about your children, mine or anybody else's, any further than we know they are saved (JD, 18:31-32, 34).[42]

This is where the matter rests. Beyond the fact that, as a group, little children are presumably heirs of the celestial kingdom, we cannot particularize as to the destiny of any one child. The Lord knows the end from the beginning. He knows how men will exercise their agency, what the consequences will be, and how their actions will affect others. He beheld the tapestry of human history long before the warp and woof of the individual lives comprising it were drawn from the skein of time. All that he does is done with but one grand objective in view: the eternal welfare and happiness of each of his sons and daughters. When we stand before him in the final judgment, all will acknowledge that "he counseleth in wisdom, and in justice, and in great mercy, over all his works" (Jacob 4:10). That is sufficient.

An Eternal Inheritance

The Prophet Nephi testified: "Behold, the Lord hath created the earth that it should be inhabited; and he hath created his children that they should possess it" (1 Ne. 17:36). Nephi is referring to the eternal, not the temporal earth. The eternal earth, being celestial, can only be possessed through obedience to the celestial law which Earth itself abides. Those attempting to possess the earth now are merely squatters on another Man's property.[43] Down through the ages wars have raged, nations have destroyed nations, and men have lied, cheated, rampaged and killed—all to possess a sliver of Earth for a sliver of time. But, as Solomon wrote, it is all "vanity of vanities," all futility. Nothing is really possessed that is not given us of the Lord, whether it be power, glory, wealth, dominion or family. The world has never learned this simple truth, therefore the world will not gain an eternal inheritance on this planet.

Some promised lands are promised forever. Those associated with this fallen earth are symbolic of, and will give way to, eternal lands of promise. Salvation is a material reality, heaven is a real place with real flowers, trees, gardens, rivers, houses, and people. In conversation with Jesus, the apostle Peter said, "Behold, we have forsaken all, and followed thee; what shall we have therefore?" In other words, we have given up our pathetic fishing boats, our miserable little houses, our pitiful possessions, and our narrow, contracted lives to follow you, now, what are we going to get out of all of this? Jesus did not rebuke Peter or promise him and his fellow apostles a cloud and a harp in some vague heaven. Instead he answered:

> Verily I say unto you, That ye which have followed me, in the regeneration [resurrection] when the Son of Man shall sit in the throne of glory, ye also shall sit upon twelve thrones, judging the twelve tribes of Israel. And every one that hath forsaken houses, or brethren, or sisters, or father, or mother, or wife, or children, or lands, for my name's sake, shall receive an hundredfold, and shall inherit everlasting life (Matt. 19:27-29).

Thrones for fishing boats! Kingdoms for hovels! A hundred things for one! A ten thousand percent increase! This is not hyperbole. To the contrary, it does not begin to indicate the "treasures in heaven" the Lord has prepared for the faithful—treasures that neither time, nature, nor man can ever take from them. Thus, the happiness they knew in the premortal estate will be multiplied "an hundredfold" when they are added upon in the resurrection with those tabernacles of flesh and bone the earth provided. And this beautiful, faithful Earth, having filled the measure of its temporal creation, will pass away to await another birth into the glorious presence of the Father and the Son.

NOTES

1 See Alma 7:20; DC 3:2; 14:7; 35:1.
2 See Moses 1:4, 38; 2 Ne. 29:9.
3 See D&C 132:19-24, 63.
4 In one of his last discourses, the Prophet Joseph Smith revealed, "God himself, the Father of us all, dwelt on an earth, the same as Jesus Christ did" (TPJS, 346; see JD, 1:356; 18:32).
5 A classic example is Jeremiah: "Before I formed thee in the belly I knew thee. . .and I ordained thee a prophet unto the nations" (Jer. 1:5). Joseph Smith was likewise foreordained (see TPJS, 365; 2 Ne. 3:7-8, 14-15).
6 His statement was in reply to a letter on the general subject from Orson F. Whitney.
7 A "doctrine of the devil" according to Joseph Smith. See TPJS, 104-05.
8 Describing God's temperament, Heber C. Kimball remarked: "I am perfectly satisfied that my Father and my God is a cheerful, pleasant, lively, and good-natured Being. Why? Because I am cheerful, pleasant, lively, and good-natured when I have His Spirit. That is one reason why I know; and another is—the Lord said, through Joseph Smith, 'I delight in a glad heart and a cheerful countenance.' That arises from the perfection of His attributes; He is a jovial, lively person, and a beautiful man (JD, 4:222).
9 Joseph Smith remarked, "The great principle of happiness consists in having a body. The devil has no [physical] body, and herein is his punishment" (TPJS, 181).
10 Brigham Young stated, "Every iota of this organization is necessary to secure for us an exaltation with the Gods" (JD, 9:286).
11 Primal intelligence and the organized spirit constitute the first two components of a resurrected being; see D&C 88:15; 93:29, 33; TPJS, 352-55.
12 Contrary to certain contemporary neo-paganistic movements, the foregoing does not mean that mankind should worship the earth. Man is the "lord of creation;" the earth is his servant and should be respected, but not deified.
13 "Rulers" suggests more than mere leadership roles in mortality; it implies kingship, Godhood. Abraham now "sitteth upon his throne" (D&C 132:29).
14 It is a doctrine of the LDS Church that all spirits entering mortality kept their first estate and were qualified for mortal birth. However, not all kept it to the same degree; all were sufficiently, but not equally worthy or prepared for mortality. Likewise, all spirits entering the second estate

do not "keep" it equally well, yet with the exception of sons of perdition, the entire human race, from purest saint to vilest sinner will eventually be saved in the kingdom of God. Relatively speaking, they will have "kept" their second estate. But what is meant by "keep"? It suggests magnifying those principles of intelligence, of light and truth, made available to us, being magnified in the things of God. The noble and great honored their first estate in a superior manner. In keeping their second estate, they will "have glory added upon their heads for ever and ever." Such ever-increasing glory characterizes those possessing exaltation and eternal lives (see D&C 131:1-4; 132:19-21, 63).

15 Joseph Smith said: "We came to this earth that we might have a body and present it pure before God in the celestial kingdom" (TPJS, 181).

16 Hinckley, 247-48. According to Abraham, the entire host of heaven destined for this earth existed before the physical planet was organized.

17 Elder Bruce R. McConkie wrote: "Men are born into mortality with the talents and abilities acquired by obedience to law in the first estate. Above all talents. . . stands the talent for spirituality" (*MM*, 234). Many factors influence and compromise the course of one's life in mortality. Some noble spirits are sent to earth in times and places and under circumstances which belie their true character and worth, much as a missionary may be sent to a difficult, undesirable field of labor. Not knowing all the facts,we are incapable of rendering a just judgment on anyone—including ourselves. The very Son of God came to earth with veiled glory; see Alma 9:26-27.

18 In observing that Moses was to stand as God to the Israelites, Joseph Smith said: "I believe those Gods that God reveals as Gods to be sons of God, and all can cry. 'Abba, Father!' Sons of God who exalt themselves to be Gods, *even from before the foundation of the world*, and are the only Gods I have a reverence for" (TPJS, 375; emphasis added).

19 The fact that all spirits are born equal in innocence before God (D&C 93:38) does not mean that all are born equal in spiritual and intellectual attainments. If they were, the foreordination of noble and great spirits would be negated and become meaningless. Parley P. Pratt observed, "although some eternal intelligences may be superior to others, and although some are more noble, and consequently are elected to fill certain useful and necessary offices for the good of others, yet the greater and the less may both be innocent, and both be justified, and be useful, each in their own capacity; if each magnify their own calling, and act in their own capacity, it is all right" (JD, 1:258).

20 Joseph Smith said, "The great principle of happiness consists in having a body. The devil has no body, and herein is his punishment" (TPJS, 181). See also Abr. 3:28; Moses 4:1-3 and Jude 6.

21 Note that Abr. 3:26 does not say that those who fail to keep the first

estate are forever denied glory; it says that they do not have glory *in the same kingdom* with those who keep their first estate. J. Reuben Clark, commenting on this passage said: "I am not undertaking to declare doctrine or Gospel, but as I read that, and as I understand it, it means that after we, so to speak, have been taken out, those who have kept their first estates, and we are not the only ones, there remains the great over-plus. They do not have the same heritage, the same kingdom, the same glory, that we shall have, and they have and can fall into the terrestrial, the telestial, and then the Doctrine and Covenants tells us there is a kingdom without any glory (D&C 88:24). My point is that we were not equal at the beginning as intelligences; we were not equal in the Grand Council; we were not equal after the Grand Council" (CR, 6 Oct. 1956, 82-84)

22 In his poetic paraphrase of Doctrine and Covenants, section 76, Joseph Smith described one class of candidates for the terrestrial kingdom as follows: "Behold, these are they that have died without law; The heathen of ages that never had hope" (*Times and Seasons*, 4 (February 1, 1843): 82-85).

23 See Rom. 2:12; 2 Ne. 9:25-27; 29:11; Alma 29:8. It is for this reason that those who "died without law" (D&C 76:72) are judged to be heirs of the terrestrial kingdom. As a group, they did not qualify for celestial glory. In connection with the Savior's advent we read: "And then shall the heathen nations be redeemed, and they that knew no law shall have part in the first resurrection; and it shall be *tolerable* for them" (D&C 45:54).

24 See 19:288; 21:198.

25 Discourse, 1 Oct. 1854, Church Archives. Orson Pratt made this same point (see JD,19:289). Jesus' own life reflects this view. Alma taught: "And he will take upon him death, that he may loose the bands of death which bind his people; and he will take upon him their infirmities, that his bowels may be filled with mercy, that he may know according to the flesh how to succor his people according to their infirmities" (Alma 7:12; see Heb. 2:18; 4:15).

26 The "cross" is not fashioned from the normal vicissitudes of life—the "slings and arrows of outrageous fortune" —which befall most of us. Instead, it is those peculiar trials which come in consequence of our commitment to Christ—the "cost" of being true to the covenants we have made.

27 This "night" is the resurrection, not physical death; see Alma 12:24; 34:33; 42:4-13.

28 See D&C 76:73; 88:99; 128:22; Isa. 24:22; 42:7; 61:1; Matt. 5:25-26; 18:34, 35; John 5:25; 1 Pet. 3:18; 4:6; *GD*, 472-476.

29 The teaching of Amulek (Alma 34:30-35) that there can be no spiritual

labor, no repentance, after death must be qualified. The dead can repent (D&C 138:58-59; TPJS, 191-92), but those who had ample opportunity in mortality, but procrastinated doing so, cannot perform a post-death labor which will qualify them for salvation in the celestial kingdom. They are saved in a lesser glory.

30 Commenting on D&C 137:10, Elder Neal A. Maxwell wrote, "There are demographics, too, to drive this doctrine: of the approximately 70 billion individuals who, up to now, have inhabited this planet, probably not more than one percent have really heard the gospel" (Maxwell, 115-16).

31 Another class of spirits that will be spared the trials of this telestial world and the temptations of Satan are those who are born during the Millennium and who are resurrected prior to the "little season" when Satan is loosed (see D&C 45:30-32, 58; 101:28-31).

32 Wilford Woodruff said, "their spirits are taken home to God" (JD, 14:126). This statements can be misinterpreted. The dead (disembodied spirits) return to God only in the sense that they are once more unembodied as they were in their first estate.

33 Orson Pratt wrote, "We see no impropriety in Jesus offering himself as an acceptable offering and sacrifice before the Father to atone for the sins of His brethren, committed, not only in the second, but also in the first estate" (*The Seer*, vol. 1, no. 4, April, 1853, 54). Having moral agency, pre-mortal spirits could and did commit sins, the violation of law. Hence, the judgment that many were "noble and great." Doubtless many sins by many spirits led to Lucifer's culminating rebellion against God. Only the Son of God could say, "I have suffered the will of the Father in all things from the beginnning" (3 Ne. 11:11).

34 See Mosiah 3:16; Moro. 8:12.

35 See Alma 5:15; D&C 29:47.

36 See D&C 29:36; 1 Ne. 17:35. Gospel covenants bring special privileges to those who honor them. For example, Joseph Smith taught that "when a seal is put upon the father and mother, it secures their posterity, so that they cannot be lost, but will be saved by virtue of the covenant of their father and mother" (TPJS, 321; see JD, 7:337).

37 See *GD*, 452. While the Lord may "take" a child through illness or accident, he is not a party to the malicious, sinful death of any child.

38 William Clayton's journal entry for 18 May 1843 records the following: "I asked the Prest.[Joseph Smith] wether children who die in infancy will grow. He answered 'No, we shall receive them precisely in the same state as they died ie no larger. They will have as much intelligence as we shall but shall always remain separate and single;'" (*WJS*, 136, n. 22).

39 The term, exaltation, is relative, not absolute. There are degrees of exaltation in mortality as well as in the life to come (see TPJS, 347).
40 Discourse, 24 Oct. 1860, Church archives; spelling corrected.
41 See JD, 4:268-69; 7:7.
42 However, Joseph Fielding Smith and Bruce R. McConkie stated children "will be heirs of exaltation" in the celestial kingdom (see *DS*, 2:54; *MD*, 675.
43 See Luke 16:10-12.

Chapter Five

Earth's Dual Creation

The Lord's course is "one eternal round," a creative cycle in which all things are brought forth with a dual, *material,* nature. He explained:

> For by the power of my Spirit created I them; yea, all things both temporal and spiritual—First spiritual, secondly temporal, which is the beginning of my work; and again, first temporal, and secondly spiritual which is the last of my work (D&C 29:31-32).

That is, all things are first organized out of spirit matter, thereafter they receive a second organization out of element, which is gross matter. In the resurrection, the two natures or organizations—spirit and element—are "inseparably connected," becoming one perfected, *spiritual* or immortal creation. This process applies to everyone and everything organized by God.

The Spirit Earth

Earth was brought forth in harmony with this principle of duality.[1] In the aforementioned doctrinal statement, "The Origin of Man," the First Presidency declared:

> The creation was two-fold—firstly spiritual, secondly temporal. This truth, also, Moses plainly taught—much more

plainly than it has come down to us in the imperfect translations of the Bible that are now in use. Therein the fact of a spiritual creation, antedating the temporal creation, is strongly implied, but *the proof of it is not so clear and conclusive as in other records* held by the Latter-day Saints to be of equal authority with the Jewish scriptures.[2]

Elder Parley P. Pratt wrote:

> The earth and other planets of a like sphere, have their inward or spiritual spheres, as well as their outward, or temporal. The one is peopled by temporal tabernacles, and the other by spirits. A vail is drawn between the one sphere and the other, whereby all objects in the spiritual sphere are rendered invisible to those in the temporal.[3]

These dual spheres are needed to accommodate mankind during the eons of time when the family of Man is divided into two groups, one being immortal spirits, the other living souls of flesh and blood. In the resurrection when spirit and flesh are united, becoming "inseparably connected" in one spiritual body, Earth's two spheres. will likewise merge into one spiritual body—the immortal home of celestial beings.

So Earth and mankind travel parallel paths. In compelling ways they share a common origin, law, and destiny, so much so that the one becomes analogous of the other. Initially organized from spirit matter, the spirit earth or world is, by nature, an immortal sphere. Being so, it is invisible to us mortals because such matter "is more refined or pure, and can only be discerned by purer eyes" (D&C 131:7). And just as the physical body is patterned after the basic image of its immortal spirit, so was this physical planet originally fashioned in the image of its spiritual counterpart.

Joseph F. Smith remarked,

> Things upon the earth, so far as they have not been perverted by wickedness, are typical of things in heaven. Heaven was the prototype of this beautiful creation when it came from the hand of the Creator, and was pronounced 'good' (JD, 23:175).

But neither Earth nor man have remained "good;" both have fallen—the former because of the latter. Still, the immortal patterns are unchanged; only the reproductions are flawed (see JD, 14:231).

The Place of Death

It is a popular, but erroneous, assumption that at death the righteous immediately return to the Father in his heavenly kingdom. This misunderstanding is based in part on Alma's statement:

> Behold, it has been made known unto me by an angel, that the spirits of all men, as soon as they are departed from this mortal body, yea, the spirits of all men, whether they be good or evil, are taken home to that God who gave them life (Alma 40:11).[4]

That is, at death the spirit returns to the same bodiless state it was in prior to mortal birth. It is in this sense that the spirit is "taken home" to God. Thus, in death, said Brigham Young, "I enjoy the presence of my heavenly Father, *by the power of his Spirit*" (JD, 17:142; emphasis added). So, in reality, the "home" to which the dead go when they die is the spirit world, the counterpart of this physical earth. Joseph Smith explained,

> Hades, the Greek, Sheol, the Hebrew, these two significations mean a world of spirits. Hades, Sheol, paradise, spirits in prison, are all one: it is a world of spirits. The righteous

and the wicked all go to the same world of spirits until the resurrection (TPJS, 310).[5]

All mankind is literally earth-bound until the resurrection.[6] Brigham Young expounded on this point:

> You read in the Bible that when the spirit leaves the body it goes to God who gave it. Now tell me where God is not, if you please; you cannot. . . . When you lay down this tabernacle where are you going? Into the spiritual world. Are you going into Abraham's bosom. No, not anywhere nigh there, but into the spirit world. Where is the spirit world? It is right here. Do the good and evil spirits go together? Yes, they do. . . . When the spirits of mankind leave their bodies, no matter whether the individual was a Prophet or the meanest person that you could find, where do they go? To the spirit world. (JD, 3:368-69).

As the human spirit dwells in and about its temporal body, so is the spirit world in close proximity to its physical counterpart. This doctrine originated with Joseph Smith:

> The spirits of the just are exalted to a greater and more glorious work; hence they are blessed in their departure to the world of spirits. Enveloped in flaming fire [glory], they are not far from us, and know and understand our thoughts, feelings, and motions, and are often pained therewith (TPJS, 326).

If the Lord would permit it, said Brigham Young, "you could see the spirits that have departed from this world, as plainly as you now see bodies with your natural eyes" (JD, 3:368).

Paradise and Outer Darkness

The Book of Mormon divides the dead into two basic groups, the righteous in a state of rest and peace called "paradise," the wicked in a state of misery and turmoil described as "outer darkness" (Alma 40:11-14). Because mankind exists on a moral and spiritual continuum, this is somewhat of an oversimplification. For the wicked and unredeemed, the spirit world is a place of relative confusion and disunity where every conceiveable notion is entertained and spiritual darkness (ignorance) prevails. Since death is like stepping through a door into another room, whatever character and personality, whatever beliefs, desires, and appetites men and women possess at death is the baggage they carry into the spirit world.[7] Consequently, the spirit world is every bit as diverse as this world. Brigham Young described the spirit world's heterogenous society:

> The spirits that dwell in these tabernacles on this earth, when they leave them, go directly into the world of spirits. What, a congregated mass of inhabitants there in spirit, mingling with each other, as they do here? Yes, brethren, they are there together, and if they associate together, and collect together in clans and in societies as they do here, it is their privilege. No doubt they yet, more or less, see, hear, converse, and have to do with each other, both good and bad. (JD, 2:137).

Nevertheless, "birds of a feather flock together" in death as in life, so that in reality the "just" are separate and apart from the wicked and are not subject to satanic temptation, power or influence.[8]

There are no children in the spirit world; all of the spirits destined for this earth achieved adulthood long before it was organized and, at birth, enter infant physical bodies. At death the spirit reenters the spirit state with the same immortal maturity it possessed prior to mortal birth. In the resurrection, the spirit reclaims the identical body, of whatever form and size, it abandoned at death and, thereafter, the body grows and develops until it attains the stature and perfection of its spirit.[9]

The righteous experience the spirit world as a realm of order and serenity where godliness thrives and truth abounds, where pain is unknown and flowers never fade. At the funeral of Jedediah M. Grant in 1856, Heber C. Kimball recounted Grant's description of the spirit world as he saw it "two nights in succession" during his last illness. Grant spoke of his dread at the thought of returning to mortality and of the perfect order and government which existed "grade after grade" among the righteous dead who were "organized in family capacities." Grant testified, "Why, it is just as brother Brigham says it is; it is just as he has told us many a time." Grant went on to describe the magnificent buildings in the spirit world, saying that the least of them was superior to Solomon's temple. The gardens, too, transcended any to be found on earth; some plants had "fifty to a hundred different colored flowers growing upon one stalk." Joseph Smith is reported to have described the spirit world as being so desireable that if men knew how magnificent it is they would commit suicide to go there.[10] It is not death, but sin we should fear. Kimball added, "If I do weep, I will weep for my own sins and not for Jedediah. If he could speak he would say, 'Weep not for me, but weep for your own sins'" (JD, 12:135-36).

Vision of the Spirit World

The climactic doctrine revealed through Joseph Smith was that mortal men and women could obtain not only immortality but divinity. In and through this apotheosis the human family is endlessly perpetuated. This is the doctrine of "eternal lives," of the "continuation of the seeds forever" (D&C 132:20-24). Those obtaining this ultimate level of salvation will, as eternal husbands and wives, honor the first commandment once more and multiply and replenish the celestialized earth with their own spirit sons and daughters. The "eternal round," having come full circle, will begin again.

Only God knows how great the "harvest" of such exalted persons will be from this one Earth, but apparently it will be vast, numbering in the millions. Their combined future spirit offspring, unlimited by any life-inhibiting factors, will be virtually numberless. So numberless that, in the words of Isaiah concerning latter-day Israel, they could well say, "The place is too strait for me; give place to me that I may dwell" (1 Ne. 21:20; Isa. 49:20). That is, their sheer numbers will oblige them to spread abroad to other worlds. Parley P. Pratt wrote:

> In order to multiply organized bodies composed of spiritual element, worlds and mansions composed of spiritual element would be necessary as a home, adapted to their existence and enjoyment. As these spiritual bodies increased in numbers, other spiritual worlds would be necessary, on which to *transplant* them (KST, 46).

That such a transplanting has occurred is indicated in President Joseph F. Smith's vision of the redemption of the dead.[11] This vision not only clarified Jesus' long misunderstood ministry in the spirit world,[12] it also greatly expanded our understanding of the vital role of that world in the plan of salvation. In the vision, Joseph F. Smith saw three groups of spirits inhabiting that world: 1) the "just" who were "awaiting the advent of the Son of God into the spirit world, to declare their redemption from the bands of death," 2) those who "died in their sins without a knowledge of the truth, or in transgression, having rejected the prophets," and 3) the Prophet Joseph Smith and other prominent latter-day Church leaders who "were also in the spirit world." That is, they were in the *same* world to which faithful elders of this dispensation "go to preach" to the unredeemed "in the great world of the spirits of the dead" (D&C 138:57-59).

The context of the vision from first to last is the spirit world; the Father's kingdom, the place of spirit birth, does not directly figure in it.[13] The one allusion to that celestial realm is found in verse 51 which states that, following Christ's resurrection, the worthy dead would "enter into his Father's kingdom, there to be crowned with immortality and eternal life" (D&C 138:51). It is at this point in the account that reference is made to Joseph Smith and other latter-day leaders who "were *also* in the spirit world:

> I observed that they were also among the noble and great ones who were chosen in the beginning to be rulers in the Church of God. Even before they were born, they, with many others, received their first lessons in the world of spirits and were prepared to come forth in the due time of the Lord (D&C 138:54-56).

However, the presence of the unborn in the spirit world is contrary to the general understanding that spirits entering mortality come directly from the Father's literal presence.[14] This belief originates in the doctrine that spirits are begotten to celestial parents in a celestial setting. It is expressed in Eliza R. Snow's hymn, "O My Father":

In thy holy habitation,
Did my spirit once reside;
In my first primeval childhood,
Was I nurtured near thy side?
(LDS Hymnal, 139).

Although it is understood that we, as spirits, did "once reside" in the Father's literal presence, in our "primeval *childhood*," nowhere in scripture does it state that we remained there.[15] While we cannot argue from silence, President Smith's account suggests the possibility that over a period of time spirits destined for this world may have left the Father's kingdom, taking up residence on the spirit Earth.

Confusion on this point stems from "spirit world" being used in Mormon discourse loosely to refer to any state in which an unembodied (pre-mortal) or disembodied (post-mortal) spirit finds itself. It has been applied to the place of spirit birth: "In the spirit world their spirits were first begotten and brought forth, and they lived their with their parents for ages before they came here" (JD, 4:216; see 8:141). And to the realm of the dead: "When he [Joseph Smith] finishes his mission in the spirit world, he will be resurrected, but he has not yet done there" (JD, 4:285). Heber C. Kimball combined the two in saying: "I have got to leave this tabernacle to go *again* into the *spirit world* to perform a work I cannot do in the flesh, that I may be prepared to receive my body *again* and enter into the *celestial world* with the Gods" (JD, 6:325; emphasis added).

As to locations, "spirit world" has been applied only to Earth's spirit world or to the Father's celestial residence. Through the years Joseph F. Smith consistently employed the expression in the former sense.[16] And in his account of the 1918 vision he made no attempt to qualify the term or its cognates.[17] Assuming he was consistent in his usage of the term throughout his account, certain conclusions may be tentatively suggested: First, both the dead and the unborn hosts are located on the spirit Earth.

Second, with so vast a population (numbering in the many billions), the spirit Earth is very likely much larger than the physical Earth. Third, spirits entering the second estate do so from the spirit Earth. This assumes that all of the unborn, as well as those actually mentioned in the vision, are to be included in the third category. If this is not the case, the unborn assigned to this world may come from either the

Father's celestial presence or from the spirit Earth—an unlikely possibility. The emphasis of the vision is on those latter-day prophets and elders who labor in the work of salvation for the dead. The absence of references to the general mass of spirits awaiting birth into mortality is, therefore, understandable.

The foregoing suggests that the journey of mankind toward eternal life involves three major steps: 1) birth and "primeval childhood" in the Father's kingdom, 2) a proving period of trial and spiritual growth on the spirit earth, and 3) the final descent into mortality, the second estate. Thus the human spirit leaves and returns to the "presence of God" from one and the same place—the spirit Earth.

Finally, the "war in heaven" was most likely fought on the spirit Earth where the character of the human family ranged from the most noble to the most rebellious.[18] For this very reason, it would seem a far more appropriate setting for a prolonged "war" between good and evil than a glorious celestial kingdom, the sacred residence of a holy Being who "cannot look upon sin with the least degree of allowance" (D&C 1:31). Then too, John's account of the conflict between God and Lucifer and the rise of the spiritual kingdom of God in pre-mortality suggests an unstable, formative state of affairs rather than a perfected condition.[19] Indeed, the "war in heaven" continues on this corrupt earth, not in any celestial setting.[20] Joseph F. Smith's vision of the redemption of the dead clarifies the long misunderstood ministry of Christ to the dead, and may be the key to a greater understanding of the role of the spirit earth in the divine scheme of things.

CREATION OF THE PHYSICAL EARTH

The Gods of Creation

Prior to organizing the physical Earth, the three members of the Godhead covenanted with each other as to their respective roles and responsibilities in the enterprise. Joseph Smith said that this covenant was everlasting and that it "relates to their dispensation of things to men on the earth; these personages, *according to Abraham's record,* are called God the first, the Creator; God the second, the Redeemer; and God the third, the witness or Testator" (TPJS, 190).[21] In harmony with the Prophet's statement, Brigham Young referred to the three personages who were directly concerned with the actual labor of creation:

> It is true that the earth was organized by *three distinct characters,* namely, Eloheim, Yahovah, and Michael, these three forming a quorum, as in all heavenly bodies, and in organizing element, perfectly represented in the Deity, as Father, Son, and Holy Ghost (JD, 1:52).[22]

That is, the three in the Creation quorum typified, but were *not identical* with, the three constituting the Godhead.

That "God the first," the Father, is called the Creator does not contradict the doctrine that the Son also bears the title Creator. The Father performs his works by and through the Son.[23] Then too, the "fullness of the Father" is the fullness of Godhood.[24] The powers exercised by those subordinate to the Father are possessed in their totality by him. Hence, in his attributes, powers, and glory he is a composite of all the Gods. In this sense he is the Creator, the Redeemer, and the Testator at one and the same time—the Most High God. He shares his powers and attributes with his associates so that their callings become extensions of his authority. He, like a Presiding High

Priest, possesses all the gifts and powers of those who serve under him even though he does not always exercise them. Thus, in employing the creative power of the Father, Jesus Christ became the Creator— "the Father of the heavens and of the earth, and all things that in them are" (Ether 4:7).[25]

"God" in the Genesis prologue— "In the beginning God created the heaven and the earth"—should be read as a collective noun. The record of Abraham explicitly states that a plurality of Gods organized Earth.[26] Abraham, in vision, was taken back through time to a point in eternity prior to "the beginning" and shown the very council in which the decision was made to organize this planet. While he does not specifically identify the Gods of creation, he writes of one who stood among the noble and the great ones who was "like unto God," who led those who were with him in organizing this planet.[27] This personage was the second ranking member of this Godhead, Jesus Christ. Speaking for the Father, he told Moses, "I am the Beginning and the End, the Almighty God; *by* mine Only Begotten *I* created these things; yea, in the beginning I created the heaven, and the earth upon which thou standest" (Moses 2:1).[28] Christ's appointment was made in that "grand council" convened for the purpose of planning the earth's organization. Joseph Smith stated:

> The head God called together the Gods and sat in grand council to bring forth the world. The grand councilors sat at the head in yonder heavens and contemplated the creation of the worlds which were created at the time. . . .In the beginning, the head of the Gods called a council of the Gods; and they came together and concocted a plan to create the world and people it (TPJS, 348-49; see also 371-72).

He also spoke of "the council in Kolob"[29] which well may have been this very council. That such was the case is not improbable since, as we have seen, Kolob has governance over this and other earths of the same order.

Matter Unorganized

The Creator said to his companions: "We will go down, for there is space there, and we will take of these materials, and we will make an earth whereon these may dwell." The materials which the Gods then organized were not created *ex nihilo*. From nothing comes nothing. The word *create* was translated from the Hebrew word *baurau*. Joseph Smith pointed out:

> Now, the word create came from the word *baurau* [or *bara*] which does not mean to create out of nothing; it means to organize; the same as a man would organize materials and build a ship. Hence, we infer that God had materials to organize the world out of chaos—chaotic matter, which is element, and in which dwells all the glory. Element had an existence from the time he had. The pure principles of element are principles which can never be destroyed; they may be organized and reorganized, but not destroyed. They had no beginning and can have no end (TPJS, 350-52).[30]

Joseph Smith reasoned that absolutely nothing could be eternal if it was created *ex nihilo*—literally out of nothing. Since matter (element) is eternal, it can be said that absolutely everything existed in "an elementary state, from eternity" (TPJS,158). Further, matter can be organized, disorganized and reorganized. Therefore, it is conceivable that at least some of this planet's "materials" were drawn from earlier creations. Indeed, the Prophet said such was the case: "This earth was organized or formed out of other planets which were broke up and remodeled and made into the one on which we live."[31] Orson Pratt, characteristically, carried this idea to the point of hyperbole:

How many thousands of millions of times the elements of our globe have been organized and disorganized; or how many millions of shapes or forms the elements have been thrown into in their successive organization and disorganization. . . is unknown to us mortals (*The Seer*, 248-49).[32]

Heber C. Kimball advanced a provocative thought, which he realized some would call "foolish philosophy," when he asked the rhetorical question: "Where did the earth come from? From its parent earths" (JD, 6:36). While this unadorned concept appears altogether naive, it has merit. Life begets life. This principle not only assures continuity from generation to generation, but from world to world as well. It is not too far-fetched, therefore, for Elder Kimball to suggest that worlds are products of other worlds. Each is a living thing and, as such, it may be organized from a "seed" or nucleus— "from its parent earths." All of the Creator's works may be literally linked one to another in an unbroken, eternal, chain.

How much debris from other planets was used in forming this small earth? And what was its consistency? Was it fine, gaseous particles, comparatively rough conglomerate, or was it massive blocks of soil and rock pressed together like a colossal snowball? Responding to this question, Orson Pratt wrote:

We are not to suppose that these elements, before they were collected, were formed into solid masses of rocks and other hard substances: and that these came rushing together—rocks being piled on rocks, breaking, crashing, and rending into millions of fragments. But no doubt through the operation of antecedent forces, *there had been a complete disorganization or dissolution* of the bodies, composed of these elements in that prior state or existence anterior to the foundation of the present globe: this being the case, the elements being separate, and apart, and widely diffused, were in a condition to come together in a state of particles, instead of aggregate masses (*The Seer*, 249).[33]

This concept of creation seems to be both reasonable and desireable, however B. H. Roberts, in an apparent attempt to reconcile Scripture with paleontology, initially favored the "aggregate masses" Pratt rejected. Roberts believed that this planet had been inhabited by pre-Adamic humanoids which, along with all other life forms, were destroyed in an earlier epoch. Earth's *fragments*, containing their fossilized remains, were then reorganized into the present planet, after which true man, together with other life forms were "brought from some other world to our own."[34]

If correct, this would serve to account for the troubling presence of supposedly pre-Adamic fossils and therefore, of death prior to the Fall. But leaving fossilized remains lying about from an earlier planet would be like planting a rose garden amidst the scattered bones of an abattoir. Such a thing would seem to be out of keeping with the perfect ways of the Creator in fashioning an *immortal* planet. In any event, in 1929 Roberts abandoned this theory, concluding that this planet is the same one that existed in pre-Adamic times. Death reigned until a global catastrophe destroyed all forms of animal life: hence, the fossils.[35] Thereafter Adam and Eve and a new animal cohort—all translated, not immortal—were placed on this earth. This theory leaves the scriptural accounts of the Creation and Fall in shreds.

Whatever the primal state of the Earth's "clay," it was miraculously fashioned upon the wheel of the divine Potter. "God spake, chaos heard, and worlds came into order by reason of the faith there was in HIM."[36] Orson Pratt sought to explain this miracle in terms of Newtonian physics:

> Jehovah spake—the elements came rushing together, not by their own power, but under the action of the self-moving forces of His Spirit. . .every particle moving toward the great common center with a resultant force, varying inversely as the square of its distance from every other particle. They

[modern physicists] would have called it the law of gravitation: while those better acquainted with the origin of the force would have called it the law by which the Spirit of God moves together the particles of matter.[37]

Whether the nascent beginnings of the planet were of long duration before it emerged from the hand of the Creator in an "empty and desolate" state on the first day is arguable. They certainly did not have to be. Nevertheless by that day, matter unorganized, infused with the Spirit of the Lord— "the law by which all things are governed"—had responded and a new Earth was born.

Place of Creation

Cosmologists theorize that about five billion years ago, our sun was formed by the congealing of a nebular cloud of gaseous materials indigenous to the Galaxy. The planets were formed from the remnants of this cloud of matter.[38] *Where* the earth was formed has much to do with *how* it was formed.

One of the startling facts taught by the Prophet and his associates is that the earth was not organized in this solar system![39] Joseph Smith prophesied: "This earth will be rolled back into the presence of God, and crowned with celestial glory" (TPJS, 181). Brigham Young using almost identical language said, "This earthly ball. . . .when it is celestialized it will go back into the presence of God, where it was first framed" (JD, 9:317).[40] Obviously, the earth could not be "rolled back" or "go back" into God's presence unless it had once been in his presence. President Young further remarked: "When the earth was framed and brought into existence and man was placed upon it, it was near the throne of our Father in Heaven" (JD, 17:143). John Taylor said it "was first organized, near the planet Kolob."[41] Kolob, according to Joseph Smith, was "nearest unto the throne of God" (see Abr., Fig. 1).

Scriptural support for this doctrine is found in the accounts of Abraham ("time") and Moses ("day") who state the planet was provided with light on the *first* time/day, but not from the sun which did not become a source of light until the *fourth* time/day. If the sequence is chronological, theoretically, this planet could be older than the *temporal* sun which did not become its source of light until *after* the Fall when the planet also became *temporal*.[42] It was rotating on its axis from its formation since night and day resulted from the presence of its *initial* light source on the first time/day of creation.

There is a universal principle of affinity or attraction which applies to all things; like does attract like, in time and eternity.[43] Spirits mingle with spirits, mortals with mortals, and resurrected beings with other resurrected beings. Any extensive or prolonged commingling of one class with another is not to be found. Indeed, the plan of salvation envisions eternal degrees of glory where men and women of similar natures are homogeneously grouped together. This principle is equally applicable to all heavenly systems.

The order of the universe observed and described by our astrophysicists is essentially different from that order to which Earth belonged in its primeval childhood. The pre-Fall planet was governed by and moved in harmony with other worlds of like nature. One senses the correctness and the appropriateness of the inspired teachings of the first three presidents of the Church. Both man and the planet on which he now dwells were brought forth as deathless creations in the presence of God. Both have fallen, becoming wanderers, "strangers and pilgrims," in a corner of an alien universe.[44]

The Two Accounts

The scriptural accounts[45] of the organization of Earth were written by Abraham and Moses between three and four thousand years ago.[46] In addition to ancient records, Abraham's remarkable account was based upon revelation through the Urim and Thummim and upon personal instruction from the Lord.[47] Moses also conversed with Jehovah face to face and received essentially identical information from him; indeed, the Lord actually dictated the account which bears Moses' name.[48] Thus, by vision and by direct, verbal instruction two witnesses were raised up to testify of Earth's divine origin.

Both Abraham and Moses were instructed on astronomy and the Creation prior to their respective encounters with the ruling powers of Egypt.[49] Being God's own witnesses of truths long shrouded in myth and fantasy, they stood unawed before the learning and sophistries of the ancient world. Each prophet-historian compliments the other while at the same time describing Earth's beginnings from his own vantage point.

Abraham allows us to look, as it were, over the shoulders of the Gods as they both planned and executed their creative labors. Moses on the other hand approaches the Creation *after the fact* when the planet was in its finished and fallen state. So with us, we think of the Creation in the past tense. This may explain why the Lord dictated the story of creation to Moses in the same sequence provided by Abraham, but without distinguishing, as Abraham does, between the organization of the planet per se and its preparation for the various life forms destined to occupy it. As Moses could plainly see, all had been accomplished— "even as I spoke." Through the instrumentality of two of God's mightiest prophet-seers, we are privileged to learn of Earth's beginnings from the perspectives of God and man, of eternity and time. It is to their inspired testimonies that we now turn.

Dual Creation Accounts?

A widely held view among Biblical scholars is that Genesis 1 and 2 contain somewhat different versions of the Creation from ancient sources erroneously combined into a single narrative.[50] Some LDS scriptorians have concluded that these chapters and their parallels in Moses and Abraham treat two creations, the first spiritual, the second temporal. This dual accounts theory stems from the fact that references to vegetation, animals, and man during the third, fifth and sixth days of creation respectively are followed by second references to them in near reverse order—man, vegetation and animals—on the seventh day.[51]

Among Latter-day Saints, Orson Pratt is the first and most notable exponent of the dual creation theory. He believed that "two creations" occurred in the one creative week: "God made the *spiritual* part of this creation during these six days' work that we read of; then he commenced the *temporal* work on the seventh day" (JD, 15:265). Referring to the first creation, he said:

> We learn, therefore, when speaking of this spiritual creation, that not only all the children of men, of all generations, and of all ages, were created spiritually in heaven, but that fish and fowls, and beast, and all animated things, having life, were first made spiritual in heaven, on the fifth and sixth days, before bodies of flesh were prepared for them on the earth; and that there was no flesh upon the earth until the morning of the seventh day (JD, 21:200).[52]

The organization of the spirit offspring of God, said Orson Pratt, constituted "the last work of creation on the sixth day."[53] He infers the physical planet was organized on the seventh day of creation:

How long he [man] had existed prior to the formation of this planet I do not know, but it is certain God seems to have formed the spiritual part of it [this planet] in the six days, and when it comes to the temporal part that seems to have been the work of the seventh day On the seventh day the Bible says that God ended his work. He did not altogether end it on the sixth, but he ended it on the seventh day (JD, 15:243).

Pratt reasoned: "I can not suppose that it would take the Lord six days to form such a little speck of a world as ours, and then for him on the fourth day to form a globe [the sun] fourteen hundred thousand times larger than the earth."[54] However, while stating that the first six days of the week concerned the *spiritual* creation, when discussing the formation of the *physical* planet he resorts to the Genesis-Moses account of events on the *first* day.

Orson Pratt's views are an admixture of the spiritual and temporal creations. The spirit earth—the organization of which he does not specifically discuss— was apparently formed the first day, the physical sun, moon, and stars the fourth day, spirit vegetation the third day, followed by spirit marine life and fowls the fifth day and land animals and man on the sixth day. The spirit organization of mankind on the sixth day is especially troubling since it requires the many billions of spirits destined for this planet to be procreated in less than a thousand years. In so severely compressing time, this scenario denies those spirits an adequate period of growth, development, and proving in their first estate.

Further, if it was the *spirit planet* that was organized on the first day of creation, there is no scriptural account of the formation of the *physical* planet which, as we shall see, was precisely the thing with which Moses was concerned. Joseph Fielding Smith did not accept the dual creation hypothesis:

> Now I repeat, *the account in Genesis one and two, is the account of the physical creation of the earth and all upon it,*

but the creation was not subject to mortal law until after the fall (*DS*, 177; emphasis in original).

The weight of scriptural evidence favors this position; the organization of the spirit Earth is yet to be revealed. All we have been told is that both mankind and the spirit Earth existed long before this physical planet came into being. The Lord has revealed what is essentially one, sequential, integrated chronicle of Earth's physical organization. The evidence suggests that the creation narratives of both Abraham and Moses pertain to events *after* the organization of the spirit earth when spirit life in all of its varieties was a *fait accompli*.

Abraham's vision of creation begins with a prologue in which the great patriarch is shown a portion of the cosmos as it existed in eternity—a descending order of spheres from immortal Kolob down to the temporal moon, sun, and earth. Their relative positions in the cosmic hierarchy are based upon their "times of reckoning" which, in turn, are declared to be analogous to the relative order of the "intelligences" over which God presides. Abraham then beheld the pre-mortal estate and its inhabitants— "the intelligences that were organized *before* the world was" (Abr. 3:22).[55] These spirits or souls were to be "added upon" with physical tabernacles. Abraham then heard one "like unto God" tell his companions:

> We will go down, for there is space there, and we will take of these materials, and we will make an earth whereon these may dwell; and we will prove them herewith, to see if they will do all things whatsoever the Lord their God shall command them (Abr. 3:24-25).[56]

That these "spirits" had already been proven in terms of the "first estate" is shown by the fact that "among all these there were many of the noble and great ones." Since their worthiness as spirits had been established, it only remained for them to demonstrate like virtue on a physical-temporal earth. Abraham's account of the creation of that "proving" Earth then follows.

Whereas Abraham figuratively stood among the Gods of the pre-mortal world when beholding the creation, Moses' vantage point was the physical Earth as it existed in his day. He described three visions given him by way of prologue to his creation account. In the first vision, Moses was shown "the world upon which he was created" together with "all the children of men which are, and which were created" (Moses 1:8).[57] The second vision was a magnification of the first. He again "beheld the earth, yea, even all of it, and there was not a particle of it which he did not behold, discerning it by the spirit of God." In like manner, he discerned each of its "numberless" inhabitants (Moses 1:27-28).

Having seen this world in its totality, Moses had a third vision in which he was shown many other lands which were also called "earth," for they, too, were inhabited. This astonishing knowledge prompted him to ask why and by what means they had been made. The answer:

> For mine own purpose have I made these things. Here is wisdom and it remaineth in me. . . .And worlds without number have I created; and I also created them for mine own purpose; and by the Son I created them, which is mine only Begotten (Moses 1:31, 33).

The Lord soon revealed that "purpose" to Moses: "this is my work and my glory—to bring to pass the immortality and eternal life of man" (Moses 1:30-33, 39). Lest Moses desire to know more concerning other earths, the Lord cautioned, "But only an account of *this* earth, and the inhabitants thereof, give I unto you." Moses responded: "Be merciful unto thy servant, O God, and tell me concerning *this earth* and the inhabitants thereof, and also the heavens, and then thy servant will be content." (Moses 1:35-36).[58] The Creator himself then made known the fundamental truths concerning the organization of this unique planet:

> And now, Moses, my son, I will speak unto thee concerning this earth upon which thou standest; and thou shalt write the things which I shall speak. An in a day when the children of men shall esteem my words as naught and take many of them from the book which thou shalt write, behold, I will raise up another like unto thee; and they shall be had again among the children of men–among as many as shall believe (Moses 1:40-41).[59]

At no point in the subsequent narrative does the Lord ever suggest that he is speaking of a spiritual creation. Indeed, if such were the case, his parenthetical explanation in Moses 3:5 (which, significantly, parallels Abraham 5:5) that *all* life was organized spiritually before any life existed "naturally" on this planet becomes a confusing redundancy.[60] Then too, both writers describe the primal earth as being immersed in water ("the deep") from which the "dry land" did not emerge until the third day. Water consists of gross element not to be found in a purely spiritual realm.[61]

Further, Abraham and Moses begin their accounts with the formation of the planet with its sea, land, and atmosphere. If they were alluding to its spiritual creation—and nowhere in their texts is this inferred—then no scriptural account of Earth's physical formation exists since no further reference to the origin of the planet as such is forthcoming from either writer. Finally, if, as Orson Pratt suggests, the physical earth was fashioned on the seventh day, then, contrary to both accounts, the "heaven and the earth" were not completed by the end of the sixth period and, therefore, the seventh day was not earth's first sabbath on which God "rested from *all* his work which he *had* made."

Age of Earth

Time is often the X factor in religion as well as science. We know something of the divine agenda, but we are less certain of its accompanying time schedule. This ignorance has led to many false predictions pertaining to the fulfillment of prophecy, especially concerning eschatology and the Second Coming.[62]

And time is equally a questionable factor where the age of this planet is concerned. We can measure present rates of radiation and sedimentation, but we do not actually know that these rates have been ever-constant. The scientific community is, therefore, not totally unified as to the validity and reliability of radioactive and allied dating methods currently employed in determining the age of rocks, fossils, and various organic objects.[63] Dating problems are further compounded by the unprecedented cataclysmic changes occurring in connection with the Fall, the Flood, and the Atonement. Sudden, violent, planet-wide events have characterized Earth's post-Fall history. And more astonishing changes lie ahead.

Even now this living planet is utterly unique among the other planets of our solar system. Indeed, as was noted earlier, some astrobiologists have concluded *it is highly questionable if there is any other planet in the known universe comparable to this watery, life-sustaining sphere.*[64] Its undeniable singularity makes it virtually impossible for anyone to establish its true age. Even estimates of the age of universe at large have proven variable. Only a few years ago its was estimated at twenty billion years, now different astrophysicists have scaled that figure down to fifteen, twelve and even eight billion years.

On this planet, we are obliged to assume a constancy in time, whereas for God time is what he wills it to be. It can be speeded up or slowed down according to his good pleasure.[65] In order to accept the divine account of Earth's organization, we must accept time as it exists for him. In this respect, it seems that this very planet may be a touchstone of faith in God.

Through Joseph Smith's translation of some of the writings of Moses in 1830 and some of Abraham's in 1842, the Lord provided additional information on the creation which not only supported, but improved upon, the Genesis account. But in the 19th century Lamarck, Darwin, and a number of geologists offered the world a new, more enlightened, version of creation—one diametrically opposite that provided by modern revelation. Was this just coincidence? The miraculous Fall of Earth to a telestial condition certainly altered its very character so as to virtually obliterate its original structure and appearance. Challenging the naturalistic views of geologists in his day, Brigham Young remarked:

> Geologists will tell us the earth has stood so many millions of years. . . .What do they know about it? Nothing in comparison. . . .I can tell them simply this—when the Lord Almighty brings forth the power of his chemistry, he can combine the elements and make a tree into a rock in one night or one day, if he chooses, or he can let it lie until it pulverises and blows to the four winds, without petrifying, just as he pleases. He brings together these elements as he sees proper, for he is the greatest chemist there is (JD, 15:126-27).

Did the Almighty use his "chemistry" to mask Earth's timeless character and appearance? Referring to its planned creation, he said, "we will prove them herewith." Is the revealed origin of Earth and mankind—which is contrary to the theories of men based upon the planet's fallen state—one of the means by which we are being proven in these last days? All things bear record of the Creator, including Earth and the revelations pertaining to it. Our attitude toward these "witnesses" is a key to the knowledge we seek: "for ye receive no witness [knowledge] until after the trial of your faith" (Ether 12:6).[66] How much we can really know about creation may be directly proportionate to how much we trust the Creator's word.

The Time-frame of Creation

How old is this organized physical planet? Scripture provides one answer, science another. The two are about 4.5 billion years apart. There is one statement in LDS literature, however, which approaches the scientific figure. It is found in a letter from William W. Phelps (a close associate of Joseph Smith) to William Smith, the Prophet's brother:

> Well, now, Brother William, when the house of Israel begin to come into the glorious mysteries of the kingdom, and find that Jesus Christ, whose goings forth, as the prophets said, have been from of old, from eternity; and that eternity, agreeably to the records found in the catacombs of Egypt, has been going on in this system, (not this world) almost two thousand five hundred and fifty five millions of years: and to know at the same time, that deists, geologists and others are trying to prove that matter must have existed hundreds of thousands of years;—it almost tempts the flesh to fly to God, or muster faith like Enoch to be translated and see and know as we are seen and known! (*Times and Seasons* 5 [1 Jan. 1844], 758).

This is the only known reference to the actual age of this eternity (a single epoch or cycle of creation, redemption, and perfection–one "round" of God's work and glory) which this earth joined only recently. It is a peculiar fact that 7000 (the number of earth years in a celestial week) multiplied by 365,000 (the number of earth days in a celestial day) equals 2,555,000,000.

In an effort to bridge the enormous gap between science and revelation on the age problem, various ideas have been advanced by religionists—all at the expense of Scripture. One suggestion is that the creation days were much longer than supposed and not necessarily of equal duration. Another is that creation was interspersed with pauses causing the overall process to be extended over a vast period of time. It is also proposed that the planet as such was produced by the same processes typical of all planets and that the scriptural account simply begins with the introduction of light on a planet already billions of years in the making.[67]

However well intentioned, from the standpoint of Scripture, these are baseless suppositions. Were it not for Abraham,[68] and Moses, we would know nothing of God's involvement in Earth's beginnings, and science would have the field to itself. But we do have these two seers and it is somewhat disingenuous to manipulate their words in order to accomdate current secular theory. Simply put, the issue is whether or not the scriptural accounts are centered in myth or reality. How we come down on this issue is a matter of faith, or the lack of it, in God, his word, and the prophets, especially Joseph Smith.

Would the Gods extend a project they had "concluded" to complete within a given time frame? Certainly lack of blueprints or materials or workers or unforseen problems would not have prompted them to do so. Nor is there any indication of work pauses by the Gods until the project was completed the seventh day when they rested from all their labors. The spiritual significance of that rest would be diminished by any pre-Sabbath Sabbaths.

As for creation beginning before the creation week, such a notion reduces the two accounts to a shambles; either this planet was organized the first day or it was not. If not, what did happen that day? Rewriting Abraham and Moses to accomodate current theory simply compromises their integrity and demeans the ways of the Lord: "Remember, remember that it is not the work of God that frustrated, but the work of men" (D&C 3:3). Starting, stopping, pausing, revising plans and schedules —this is what men, not Gods, do.

And why would the Almighty take billions of years to do what he could do virtually in an instant if he so chose? The prophecies pertaining to the future destiny of Earth, together with the historical accounts of his dealings with man, certainly do not support such a notion. The Gods did not even begin to organize the physical Earth until after the entire human race had been begotten as spirits in the pre-mortal world.[69] They are not neophytes in the business of creation. Did they takes billions of years to organize each of those billions of worlds Moses speaks of? Is it reasonable to assume that billions of spirits were kept waiting billions of years while the Gods who were sent down to organize this planet cast their priesthood powers aside and deferred to the forces of nature over which Priesthood rules?

As we have seen the Almighty works by the power inherent in his spoken word: "I am the same which spake, and the world was made, and all things came by me" (D&C 38:3; see Jacob 4:9). The Gods are miracle workers. Can finite man ever hope to understand their science? Moroni asked: "Who shall say that it was not a miracle that by his word the heaven and the earth should be?" (Morm. 9:17). The miracles of Jesus prove that God is bound neither by time nor by law as we mortals experience time and law.[70] For example, the natural processes involved in wine production require many months of planting, growing, harvesting, pressing, and aging. Yet Jesus, employing higher laws of organization instantaneously produced wine which was superior to the natural product.[71]

Modern science is also able to speed up the processes of nature—both creative and destructive—and quickly accomplish what would ordinarily take many years. In a thousand ways, this planet is subdued, modified, transformed, and utilized to produce new products and conditions in a micro-fraction of the time any conceivable "natural" process might entail. So, too, the time required to bring this planet with all of its myriad life forms into existence would depend entirely upon the processes employed.

The Six Times of Creation

The Creation Days

Genesis literalists maintain that the entire creation enterprise was accomplished in six, twenty-four hour days.[72] But Moses does not tell us how long the days were and there is nothing in his overall text to suggest their duration. Hence, Brigham Young's comment:

> It is said in this book (the Bible) that God made the earth in six days. This is a mere term, but it matters not whether it took six days, six months, six years, or six thousand years. The creation occupied certain periods of time. We are not authorized to say what the duration of these days was, whether Moses penned these words as we have them, or whether the translators of the Bible have given their words their intended meaning (JD,18:231).

George Q. Cannon remarked:

> Joseph taught that a day with God was not the twenty-four hours of our day; but that the six days of creation were six periods of the Lord's time (JD, 24:61).[73]

The yardstick of time lacking in Moses' account is provided by Abraham. He informs us that Kolob is a world having the same measurement of time as that of the celestial residence of God himself.[74] We read in Abraham:

> And the Lord said unto me, by the Urim and Thummim, that Kolob was after the manner of the Lord, according to its times and seasons in the revolutions thereof; that one revolution was a day unto the Lord, after his manner of reckoning, it being one thousand years according to the time appointed unto that [earth] whereon thou standest (Abr. 3:4).[75]

Abraham equates a "day" on the newly-formed earth to a "time" on Kolob:

> And it came to pass that it was from evening until morning that they called night; and it came to pass that it was from morning until evening that they called day; and this was the second time that they called night and day (Abr. 4:4; see vrs. 5, 8, 13, 19, 23, 31).

Note that he consistently employs the terms "day" and "night" in reference to the presence or absence of light, not to a specific period of time as such. Only when combined does a day and a night equal a time. Moses describes each day in Hebraic terms— "the evening and the morning."[76] But Abraham characterizes them as both days and times. First he tells us that "from evening until morning" was Earth's period of darkness and that "from morning until evening" was its period of light. He then states that each cycle of darkness and light equaled a "day" or a "time." Thus, he uses the two terms to distinguish between a "day" on Earth and a "time" on Kolob; for while they were equal in duration, they were dissimilar in nature. Whereas a "day" on Earth was characterized by alternating periods of darkness and light, a "time" on Kolob knew no darkness "from evening until morning" or "from morning until evening."

Joseph Smith's previously cited poetic paraphrase of "The Vision" (D&C 76), suggests that the "council in Kolob" and the "grand council" which was convened to plan the organization of Earth are one and the same. The Gods would logically schedule their labors relative to their own time frame. (What other time frame did they have prior to earth's organization?) And they did. Abraham, tells us that the Architects of this world planned it according to Kolob's measurement of time. It was to be a one week project to be begun and completed within seven thousand earth years.[77] However, the Gods did not need thousands or even hundreds of years to achieve their purposes. Each day's labors were almost surely accomplished in a fracton of the allotted time.[78] We shall see that Earth's creation was deliberately programmed to take place over a week of celestial time as part of a larger design for it and mankind.

Abraham recounts the various phases of creation in terms of first, second, third, etc., celestial times. This is clear from his consistent and singular usage of the word "time" throughout his account. Nowhere does he attempt to redefine the term. Indeed, he continues to refer to "time" in the same way even after concluding his account of the creation.

In writing of subsequent events, Abraham tells us that Adam was warned that he would die "in the time" in which he ate the forbidden fruit.[79] To emphasize that time had not been redefined, Abraham comments: "Now I, Abraham, saw that it was after the Lord's time, which was after the time of Kolob; for as yet the Gods had not appointed unto Adam his reckoning" (Abr. 5:13).[80] God's time, celestial time, Kolobian time, prevailed throughout the six days of creation and thereafter during the sojourn of Adam and Eve in the Garden of Eden. There is no basis for the argument that Adam was subject to one measurement of time while the planet on which he lived was subject to another.

Summary of Earth's Organization

Some exegetes interpret Abraham as only describing the "plan" or "blueprint" of creation while Moses provides the actual implementation of that plan. Not so. Abraham summarizes the labors of the Gods in organizing and preparing the earth for life as accomplished actions: "It was so, even as they ordered" (4:7, 9, 11); "they were obeyed" (4:10, 12); they "watched those things which they had ordered until they obeyed" (4:18). Abraham 4:1-25 treats the completed organization and preparation of the planet for life. Abraham 4:26-5:3

summarizes a subsequent council of the Gods dealing with placing human and animal life on the lifeless planet. Abraham 5:4-20 describes the actual coming of that life to the planet.

The word "plan" appears but once in Abraham's account where it is used in the context of an *on-going* enterprise. Like architects overseeing the construction of a mighty edifice, "the Gods saw that they *would be* obeyed, and that their *plan* was good" (Abr. 4:20).[81] When the textual unity of Abraham's record is maintained, it is apparent that he, like Moses, is telling us what the Gods did, not simply what they planned to do.

Moses' account is generally interpreted as stating that Earth was not only prepared for, but supplied with, biological life beginning with vegetation on the third day. This assumption stems from the use by Moses of such clauses as: "it was done as I spake" (2:5); "it was so, even as I spake" (2:6, 7, 11, 30); "it was done" (2:6); and "it was so" (2:9, 15, 24, 26). Hence the belief that the things spoken of on a given creation day were actually *accomplished* on that day, rather than that God simply willed them to be done and *confirmed* that eventually they would be done. For example, Moses tells us that on the sixth day *before* any life was placed on Earth, the Father decreed that animals should eat "every clean herb for meat; and *it was so even as I spake*" —meaning that when finally brought to Earth, all animals were herbivorous as God previously ordained (Moses 2:30).[82]

Unlike Abraham, Moses does not distinguish between the planning and preparation for life and its actual placement on Earth. He combines the two. This leads to the erroneous conclusion that what the Mosiac account says was "done" on a given day was, therefore, *completed* on that day. However, it is plain from Moses 3:5 that Moses understood that no living thing was found on this planet until the seventh day of creation.

From the Lord's standpoint, the scriptural account is a child's version of creation. Nevertheless it contains eternal truths with profound implications.[83] The organization of the earth was a miracle. No scientific theory on the origin of planets can account for the manner in which it was organized. It did not condense from a thundering cacophony of cosmic debris, of falling ice and rocks. It did not become a global hell of sulphurous gases and swirling lava. Instead, the Creator brought it forth without the sound of hammer or chisel as a *watery* sphere, an endless, primordial sea soaring through the heavens. No winds blew across those dark waters, no sound broke the awesome silence which enveloped them like a shroud: "darkness reigned upon the face of the deep, and the Spirit of the Gods was brooding upon the face of the waters" (Abr. 4:2, 6).[84] Then the enveloping darkness[85] vanished away as Earth's prevailing waters were bathed in light eminating, not from the sun, but from a glorious, immortal sphere. The nascent planet began to rotate on its axis in harmony with the rhythm of Kolob, introducing Night and Day. The first time/day of creation ended.

The second time/day involved the heavens, two contiguous, but distinct, regions of inner and outer space. The nearest, most proximate, to the planet of these heavens is the "firmament" or "expanse"[86] —the biosphere, the realm of winged creatures and of that life-sustaining atmosphere which now envelops the planet with diminishing density from the troposphere to the ionosphere. Beyond this atmospheric region lies the outer expanse of heaven encompassing not only the sun and the moon, but the sidereal, starry, heavens as well. Although the sun and moon are introduced on the fourth day, the myriad stars in the sideral heavens are mentioned only parenthetically ("and the stars also were made even according to my word") to complete the overall picture.[87] The Sun and the moon did not shine upon the earth until after the Fall on the seventh day.[88]

On the third time/day, the Gods spoke and "dry land" rose up out of the global deep.[89] Earth's three spheres of life–air, sea, and land–were now in place. The Gods then prepared the dry ground and the waters for the various orders of life that would subsequently inhabit them. This was necessary because the planet was probably devoid of even micro-organisms when first formed. It had to be readied for planting, much as a farmer harrows and fertilizes the ground before seeding, and for marine life much as an aquarium is readied for fish. The preparatory work may have included putting dormant seed of all kinds in the dry ground which did not germinate until the seventh day when a mist watered the barren ground. Brigham Young asked:

> Shall I say that the seeds of vegetables were planted here by the Characters that framed and built this world–that the seeds of every plant composing the vegetable kingdom were brought from another world? This would be news to many of you (JD, 7:285).[90]

Thus, on the third, fifth, and sixth times/days, respectively, the Gods prepared the soil for vegetation, the waters for marine life, the heavens for fowl, and the ground for land animals and human life. In every instance, the Gods were either obeyed or saw that they would be obeyed.[91]

The Second Council

As the sixth time /day drew to a close, the Gods again "took counsel" to organize man, male and female, in their own image, *and* to organize "every living thing that moveth upon the earth."[92] This reaffirmation was stated in terms of intent, not of completion: "We *will* bless them. We *will* cause them to be fruitful. We *will* give them every green herb, etc." The sixth time ended with the Gods saying, "We *will* do everything that we have said, and organize them; and behold, they *shall* be very obedient. . . . And thus we *will* finish the heavens and the earth, and all the hosts of them" (Abr. 4:31-5:1).[93] Making and finishing are two different things. The Gods *made* (organized) the earth during the first six times/days, by forming the planet and preparing it for the myriad life forms destined to dwell upon it. They *finished* their work, not by further creative activity, but by introducing those life forms on the planet. When the sixth time/day ended, Earth was a magnificent, but empty, mansion—without furnishings or occupants.

The Seventh Time/Day

So when Earth's first Sabbath—the seventh day of creation—dawned, the planet was a silent, lifeless world. It had been "prepared" for life, but there was no living thing upon it. So to fulfill their creative "day," the Gods brought life to it.[94] Abraham wrote, "For the Gods *had not caused it to rain* upon the earth when they counseled to do them [complete their creative plan], and *had not formed a man* to till the ground" (Abr. 5:5). Every form of life existed, but not on this planet. Then on the morning of the seventh day of creation, the rains came and sleeping seeds responded, bringing forth grasses, plants, flowers and trees to bless and sustain that awaiting life with "the good things which come of the earth" (D&C 59:17).[95] The Garden of Eden came into being.

In Moses' account, the Lord confirms and elaborates on Abraham's words:

> For I, the Lord God, created all things of which I have spoken, *spiritually,* before they were *naturally* upon the face of the earth. For I, the Lord God, *had not caused it to rain* upon the face of the earth. And I, the Lord God, had created all the children of men; and not yet a man to till the ground; for in heaven created I them; and *there was not yet flesh upon the earth, neither in the water, neither in the air;* But I, the Lord God, spake, and there went up a mist from the earth, and watered the whole face of the ground. And I, the Lord God, formed man from the dust of the ground, and breathed into his nostrils the breath of life; and man became a living

soul, the *first flesh* upon the earth, the *first man* also (Moses 3:5-7; see Abr. 5:5-7).

In language which cannot be gainsaid the Creator himself declares man to be "the first flesh upon the earth"—no other living species *of any kind* was to be found on it. Nor is the Lord referring to *generic* man; he is speaking of one solitary soul—the man Adam, an immortal human being who remained "alone" until the Gods provided him with a companion "meet" for him—Eve.[96]

It was this immortal woman, not any creature, whom Adam declared to be "bone of my bones, and *flesh of my flesh.*" At no time in either creation narrative is the word "flesh" equated with mortality, for mortality did not exist on this planet in the beginning.[97] A mortal man and woman could not bring to pass the plan of salvation. How could Adam and Eve fall if they were already fallen? No living thing on this planet was mortal at that time. Orson Pratt provides a very comprehensive doctrinal statement on this point:

> Man, when he was first placed upon this earth, was an immortal being, capable of eternal endurance; his flesh and bones, as well as his spirit, were immortal and eternal in their nature; and it was just so with all the inferior creation—the lion, the leopard, the kid, and the cow; it was so with the feathered tribes of creation, as well as those that swim in the vast ocean of waters; all were immortal and eternal in their nature; and the earth itself, as a living being, was immortal and eternal in its nature (JD, 1:281).

Joseph Fielding Smith agreed: "Now when Adam was in the Garden of Eden, he was not subject to death. There was no blood in his body and he could have remained there forever. This is true of all the other creations."[98]

SEQUENCE OF CREATION

Abraham	Moses (Genesis)	Time/Day	Event
		Earth is prepared for life.	
4:1-2	2:1-2	First	1. Heavens and Earth organized ...global "deep" covers planet
4:3-5	2:3-5	First	2. Light provided....Day and Night begin
4:6-8	2:6-8	Second	3. Atmospheric heavens formed between lower and upper "waters"
4:9-10	2:9-10	Third	4. Primal continent emerges from global sea
4:11-13	2:11-13	Third	5. Land prepared for vegetation
4:14-19	2:14-19	Fourth	6. Sun, Moon, stars to serve the earth
4:20-23	2:20-23	Fifth	7. Waters prepared for marine life; lower heavens for fowl
4:24-25	2:24-25	Sixth	8. Earth prepared for land creatures
4:26-28	2:26-28	Sixth	9. Gods council to organize man —male and female
4:29-30; 5:9	2:29-30; 3:9	Sixth	Man and animals to be herbivorous
		Life Comes to Earth!	
5:6	3:6	Seventh	10. A mist waters the dry ground
5:7-8	3:7-8	Seventh	11. Adam—the "first flesh"—
5:6, 8-10	3:6, 8-14	Seventh	12. Vegetation comes forth— the Garden in Eden
5:14-19	3:18, 21-25	Seventh	13. Woman—Eve—*
5:20-21	3:19-20	Seventh	14. Animal kingdom

*As with Adam, Eve's creation is symbolic. Whether "Woman" came to Earth before or after the various species of animals is of little moment. However, both accounts may be correct: "Woman" was begotten before the animals (Abraham) but not "brought" to Adam until after they were named (Moses).

Procreation of Man

The culmination of creation took place on the seventh day: "God made the world in six days, and on the *seventh* day he finished his work, and sanctified it, and also *formed man* out of the dust of the earth" (D&C 77:12).[99] The scriptural account of the origin of Adam and Eve should not be interpreted literally: Adam was not molded from "the dust of the ground" nor was Eve formed from one of his ribs. The account is symbolic of the sacredness of the sexual union between man and woman and of the oneness which should exist between husband and wife: "bone of my bones and flesh of my flesh" (Moses 3:23-24).

Brigham Young understood the symbolism:

> When you tell me that father Adam was made as we make adobies from the earth you tell me what I deem an idle tale. When you tell me that the beasts of the field were produced in that manner, you are speaking idle words devoid of meaning. There is no such thing in all the eternities where the Gods dwell. Mankind are here because they are the offspring of parents who were brought here from another planet, and power was given them to propagate their species, and they were commanded to multiply and replenish the earth (JD, 7:285).

He further said:

> He [God] created man, as we create our children; for there is no other process of creation in heaven, on the earth, in the earth, or under the earth, or in all the eternities, that is, that were, or that ever will be (JD 11:122).

This process is operative in both time and eternity. Indeed, it is LDS doctrine that the human family was spiritually procreated by divine Parents. The previously quoted declaration of the First Presidency states:

> All men and women are in the similitude of the universal Father and Mother, and are literally the sons and daughters of Deity. . . .man, as a spirit, was begotten and born of heavenly parents, and reared to maturity in the eternal mansions of the Father, prior to coming upon the earth in a temporal body to undergo an experience in mortality.

President Joseph F. Smith, who issued this statement, was even more emphatic in public discourse:

> Man was born of woman; Christ, the Savior, was born of woman and God, the Father, was born of woman. Adam, our earthly parent was also born of woman into this world, the same as Jesus and you and I.[100]

This doctrine originated with the Prophet Joseph Smith who asked,

> Where was there ever a son without a father? And where was there ever a father without first being a son? Whenever did a tree or or anything spring into existence without a progenitor? And everything comes in this way (TPJS, 373).

Earth's Two Sabbaths

Did the Gods need six thousand years to organize Earth? Why, then, did their labors extend over six celestial days? And why did the Lord's accounts through Abraham and Moses distinguish between the labors of the Gods and the rest of the Gods? One obvious answer is that they did so to establish the law of labor and rest on the newly formed planet, the law which reflects the dual nature and, therefore, the dual needs of us mortals. The Gods taught mankind the law of the Sabbath, first by example in their creation labors, and then by precept in their written word:

> Remember the Sabbath day, to keep it holy. Six days shalt thou labour, and do all thy work. But the seventh day is the the Sabbath of the Lord thy God: in it thou shalt not do any work.... .For in six days the Lord made heaven and earth, the sea, and all that in them is, and rested the seventh day: wherefore the Lord blessed the Sabbath day, and hallowed it (Ex 20:9-11; see D&C 59:9-13).

In honoring the Sabbath, we render to the body what is the body's, and to the spirit what is the spirit's. We testify of the need for both physical activity and spiritual rest—of the difference between the temporal bread of the flesh, and the spiritual bread of the spirit.[101] Then too, the Sabbath is a continual reminder of the creation and of the fact that "The earth is the Lord's, and the fullness thereof; the world, and they that dwell therein" (Ps. 24:1). We are but steward's of God's property and we will be obliged to give an accounting of our stewardship over it. If mankind really acknowledged this fact, how different would be our treatment of Earth, and of one another.

More germaine to this book is the fact that the two accounts of the Creation, as is shown by this previously quoted revelation, are prophetic:

> We are to understand that as God made the world in six days, and on the the seventh day he finished his work, and sanctified it, and also formed man out of the dust of the earth, even so, in the beginning of the seventh thousand years will the Lord sanctify the earth, and complete the salvation of man (D&C 77:12).[102]

That is, it was ordained that Earth's seven days of organization and preparation should be followed by another seven days of temporal existence.

The spiritual and temporal stages of Earth's career are closely identified with the Son of Man whose labors as Creator prior to and during Earth's first Sabbath find fulfillment in his rule as King over the Kingdom of God during Earth's second Sabbath. This is why the Prophet Joseph Smith, alluding to Christ's millennial reign, wrote "the Sabbath of creation" will be "crowned with peace" (TPJS, 13). Both Sabbaths are the Lord's day, both introduce a new creation, and both are approximately a thousand years in length.[103]

These parallel weeks indicate that the pivotal events pertaining to Earth were foreordained with prophetic symmetry as a divine chiasmas, the central point being the atonement and resurrection of Jesus Christ in the meridian of time. Therefore, the major prophetic milestones from the beginning the world to the first coming of Christ are to be passed again, but in an inverse, modified order. All this, from first to last, and from last to first, within a time frame of two celestial weeks, or approximately fourteen thousand years.[104]

While every Sabbath is made to bless man, the first was made to receive him. Through the veil of forgetfulness came Adam, the lord of creation, together with his eternal helpmeet, Eve. They were followed by a retinue of subjects—the myriad forms of life comprising the generations of the heavens and Earth. And so mankind's second estate began in a deathless garden, in paradisiacal glory—the glory with which Earth will again be enveloped when the Lord of the Sabbath returns to sanctify and re-endow it with the splendor it once knew.[105] For in the beginning, Earth was a most peculiar planet—a physical-spiritual realm soaring among eternal worlds in the presence of the Gods.

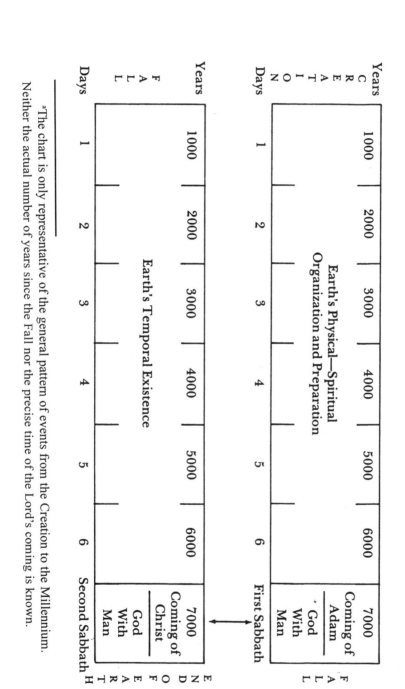

NOTES

1 See D&C 93:35; 101:25; 131:7.
2 *The Improvement Era,* 13: (November, 1909), 77; emphasis added ; see JD, 18:243.
3 *Key to the Science of Theology*, 1st ed. (Liverpool, 1855), 126-27. The spatial relationship of the spirit earth to the physical earth is unrevealed.
4 See Eccl. 12:7. Orson Pratt mistakenly taught that the dead return to God's literal presence after which the wicked are cast out (see JD, 2:238).
5 The parable of Lazarus and the Rich Man (Luke 16:19-31) confirms the Prophet's statement (see JD, 3:94-96; 11:126). Only after the resurrection do the celestial dead permanently return to the Father (see JD, 3:113, 369; 7:173; 8:28-29). Heber C. Kimball said the saints will be earthbound long after the resurrection (see JD, 1:356).
6 Heber C. Kimball noted, "I never shall come into the presence of my Father and God until I have received my resurrected body, neither will any other person" (JD, 3:113).
7 According to Oliver B. Huntington, Joseph Smith said, "A man can do as much in this life in one year as he can in ten years in the spirit world without the body" (Andrus, *They Knew the Prophet*, 61; see Hinckley, 241).
8 See JD, 3:371; 4:134; 7:174.
9 See TPJS, 200; GD, 455-57. There is confusion on this point due to somewhat contradictory reports of Joseph Smith's teachings on the subject (see WJS, 136, n. 22; HC, 4:553-57).
10 Under the date of 19 Aug. 1877, Charles L. Walker recorded the following: "He [Wilford Woodruff] referred to a saying of Joseph Smith, which he heard him utter (like this) That if the People knew what was behind the vail, they would try by every means to commit suicide that they might get there, but the Lord in his wisdom had implanted the fear of death in every person that they might cling to life and thus accomplish the designs of their creator." (*Diary of Charles Lowell Walker*, 2 vols., A. Karl and Katharine Miles Larson, eds., [Logan: Utah State University Press, 1980], 1:465-66. See JD, 14:231; 17:442; 26:192-93.
11 The vision was received 3 Oct. 1918, accepted by the leading Church councils on 31 October. President Smith died six weeks later on 19 Nov. It was added to the Doctrine and Covenants as section 138 in 1981.
12 In accepting the somewhat incomplete statement in 1 Pet. 3:18-20, early leaders of the Church, including Joseph F, Smith, taught that

Jesus personally preached to the unsaved during his sojourn in the spirit world (see TPJS,. 219, 222; JD, 3:96; 13:76; 18:91-92; 26:82-83).

13 Referring to the celestial world, Joseph Smith explained: "There are two kinds of beings in heaven: 'resurrected personages' and yet-to-be resurrected 'spirits of just men made perfect'" (see D&C 129:1-2). The world described by Joseph F. Smith was exclusively a world of spirits.

14 Although this concept is well-founded in Mormonism, there is no explicit scriptural statement to this effect. It is extrapolated from such passages as D&C 76:24 and Abr. 3:22-23, and from statements by Joseph Smith, Brigham Young and other LDS leaders (see TPJS, 373, 375; JD, 4:182, 217-18, 268; 18:290-91).

15 Just as the spirits of the dead in entering the spirit world "are taken home" to God, so, too, the spirits of the unborn, though on another spiritual sphere, are still at "home" with God. Spirits by their very nature never leave his presence. See Andrus, *God, Man, and the Universe*, 187-88.

16 See JD, 18:92, 274; GD, 460-61, 469-72.

17 See D&C 138:16, 54, 56-57.

18 There was "war in heaven," but *what* heaven? See JD, 18:294. The Greek word for heaven, *ouranos*, has several connotations and literally means something airy, cloud-like or etherial, as opposed to something solid and earth-like. As for the Father and the Son appearing in the spirit world in connection with the "war," there is no reason why they could not. The Son was a spirit at that time, his Father a resurrected being. The spirits of the dead can "mingle with Gods" and may behold the Father and the Son (see JD, 14:231; 17:143; Brian H. Stuy, *Collected Discourses*, 1:77; 2:49).

19 See JST, Rev. 12:1-11.

20 See D&C 76:28-29.

21 The Prophet also said: "Any person that has seen the heavens opened knows that there are three personages in the heavens who hold the keys of power, and one presides over all" (TPJS, 312); see 370. While there is but one true God, a plurality of deities share the title (see D&C 20:28; Alma 11:26-44; John 17:20-23; 1 Cor. 8:5-6; Col. 2:9).

22 Upon coming to Earth, Michael, the archangel, assumed the name, Adam (see D&C 27:11; 107:54; 128:21). Unlike Eve, there is no scriptural account of the naming of Adam.

23 See Moses 1:33; D&C 93:8-10; Col. 1:16.

24 See Col. 1:19; 2:9; D&C 93:12-16.

25 See Mosiah 15:1-4; Talmage, *Articles of Faith,* appendix two, 465-67; *Jesus the Christ,* 33. Speaking of those who become the sons of God, John Taylor said, "they, through the atonement might be exalted, by obedience to the law of the Gospel; *to the Godhead*" (*Mediation and Atonement,* 133).

26 See Abr. 4, 5. A plurality of Gods is also indicated in Gen. 1:26: "And God said, Let us make man in our image." See Moses 2:26 which indicates that the Father was speaking to his Only Begotten (see John 1:14, 18; 1 John 4:9).
27 See Abr. 3:23-24.
28 That the Only Begotten was the chief organizer of this planet is so well attested in both ancient and modern scripture as to make extended discussion unnecessary (see Moses 1:32, 33; D&C 38:13; 76:23, 24; John 1:1-14; Col. 1:16).
29 See "A Vision" (a paraphrasing of D&C 76), *Times and Seasons*, 4 (Feb. 1843), 82-85.
30 See also TPJS, 181; JD, 12:238; 19:286; 24:61; 26:27.
31 *The Words of Joseph Smith*, 60. See Richards, 287.
32 See JD, 16:316; 18:293, 294; 20:274; 21:322.
33 See JD, 13:248; 21:32.
34 *Man's Relationship to Deity*, 279-82.
35 See Roberts, *The Truth, The Way, the Life,* chaps. 31-32. James E. Talmage also believed that death reigned on the earth at the animal level prior to the coming of Adam and Eve. See his address, "The Earth and Man," (9 Aug. 1931, Salt Lake Tabernacle.) It appeared in several Church publications including *The Deseret News*, 21 Nov. 1931.
36 Lectures on Faith, no. 1, par. 22.
37 Orson Pratt lectured extensively on the subject of astronomy at the Deseret University (now the University of Utah). His lectures were compiled and published in 1937 by Nels B. Lundwall under the title *Wonders of the Universe.*
38 Called the Condensation Theory, it was preceded by the theory that the planets were formed as a result of a collision between our sun and another star. In 1988 a prominent science historian wrote: "It is one of the humbling truths of contemporary science that, while we theorize about the origin of the universe, we do not yet fully understand how our own little planetary system began." Ferris, 167.
39 Orson Pratt assumed that Earth was created in this solar system. See JD, 16:315.
40 See JD, 7:163; 8:8; Words of Joseph Smith, 84.
41 John Taylor, "Origin and Destiny of Woman," *The Mormon* [New York] (August 29, 1857), as quoted in Lundwall, *The Vision*, 146.
42 See Gen. 1:2-13; Moses 2:2-13; Abr. 4:2-13.
43 See D&C 88:40.
44 The word planet is taken from the Greek planetes meaning, wanderer.
45 The creation sequence in the LDS Endowment will not be considered. It differs somewhat from the scriptural accounts being essentially a dramatic and symbolic, rather than historical, treatment of the subject.

Brigham Young declared: "We admit the history that Moses gives of the creation or organization of this earth, as stated in his writings, to be correct" (CR, 8 Oct. 1875). Joseph F. Smith agreed: "The Bible account, being the most rational and indeed *only historical one of the creation* and the dealings of God with men, we are constrained to accept it, in the main, as truth" (JD, 15:326; emphasis added).

46 Abraham dates from about B.C. 2000, Mose, B.C. 1300.
47 Abraham's initial knowledge of creation was obtained from the records of the prediluvian patriarchs: "Therefore a knowledge of the beginning of the creation, and also of the planets, and of the stars, as they were made known unto the fathers, have I kept even unto this day" (Abr. 1:31; see 3:1, 11, 15, 22).
48 The second and third chapters of the writings of Moses (as found in the Pearl of Great Price) are an expanded, but otherwise parallel, version of Genesis 1 and 2.
49 See Abr. 3:15; Moses 1:17, 26.
50 The original texts of ancient scriptures had no chapter divisions or versification. It is especially important to ignore chapter breaks when studying the creation accounts.
51 Gen. 1 and 2 is paralleled verse for verse in Moses 2, 3 and, to a lesser extent, by Abr.4, 5.
52 See 16:316-18; 18:327.
53 JD, 15:244.
54 The Sun is 10 million times larger, and its mass 333,000 times greater, than that of Earth.
55 Since Abraham saw that the spirits of men were already in existence before the creative week began, Orson Pratt's view that man was spiritually organized on the sixth day of that week appears untenable (see JD 21:200).
56 That the entire human race existed as spirits before the physical planet was formed is also supported by a doctrinal statement of the First Presidency. See Moses 3:4-5; D&C 38:1; 49:16-17. See JD, 21:290.
57 Moses seems to have seen the spirit Earth as well as the physical Earth.
58 Moses was seemingly unaware of the existence of Abraham's creation narrative; otherwise he would have known how and why this earth was organized.
59 This revelation to Moses marks the beginning of his writings in the Old Testament. The fact that God, not Moses or some unknown scribe, is the author of the Biblical account of the creation has been lost from the Genesis account. The same basic creation narrative, with significant additions, was revealed to Joseph Smith in 1830. In it, the first person "I, God," is used rather than the third person "God said," of Genesis.
60 Note that these verses refer only to the spirit origin of the various life

forms, not to the earth per se. Joseph Fielding Smith said, "The statements in Moses 3:5 and Genesis 2:5 are interpolations thrown into the account of the physical creation, explaining that all things were first created in the spirit existence in heaven before they were placed on this earth" (*Doctrines of Salvation*, 1:76; emphasis in original).

61 Vicarious baptism for the dead is an earthly ordinance which is made necessary by this very fact. See D&C 128:12-14.
62 See Mark 13:32; D&C 49:7; 130:14-17; 133:11.
63 For a conservative treatment of the dating problem see Cook, *Science and Mormonism*, and *Scientific Prehistory*.
64 This is the thesis of some researchers in the newly merging science of astrobiology. See, Ward & Brownlee, *Rare Earth*.
65 Einstein's general theory of relativity demonstrates the variableness of time under different conditions.
66 While this suggestion is undoubtedly laughable to many, the fact remains that the most direct assaults on the validity of Scripture have centered in those theories concerning the origin and age of earth and man. The current retreat from belief in divine moral law and the validity of Scripture is a direct consequence of these assaults.
67 This possibility was expressed by Charles W. Penrose who felt that geology had "demonstrated the fallacy of the idea that the earth is such a young planet in the universe." He added: "How many ages upon ages passed from the time called 'in the beginning,' to that when God called forth the light out of the midst of the darkness, cannot be gleaned from any revelation or scripture ancient or modern, that is now known to man" (*The Improvement Era* [May, 1909], p. 507.) In addition to Penrose, other prominent LDS leaders who have sought an acceptable compromise with the scientific point of view were B. H. Roberts, James E. Talmage, and John A. Widtsoe.
68 On several occasions, Joseph Smith made reference to the knowledge he had gained concerning God and his creations while translating the records of Joseph and Abraham. See TPJS, 118, 190; 251, 373.
69 See D&C 49:17; Abr. 3:22; Moses 3:5.
70 This may be one reason why God must be excluded from scientific discourse. Science is obliged to assume consistent operations of natural law for its own validity and reliability. But miracles are, by definition, modifications of natural law which render science mute.
71 See John 2:1-10.
72 They are naive because they assume, not having or accepting Abraham's record, that the twenty-four hour day existed from the first day of creation even though the Sun is not mentioned until the fourth day and the Fall had not occurred.
73 Orson Pratt acknowledged that, while the specific length of time

required to perform the actual work of creation was unknown, the overall period was six thousand years. See JD, 14:234-35; 16:317.
74. Joseph Smith taught: "In answer to the question—Is not the reckoning of God's time, angel's time, prophet's time, and man's time, according to the planet on which they reside? I answer, Yes" (D&C 130:4-5). The Creator measures time according to the rotation of his celestial world, even as we measure time according to the rotation of this planet.
75. Joseph Smith revealed: "One day in Kolob is equal to a thousand years according to the measurement of this earth" (Facsimile 2, figure 1, The Book of Abraham). Knowledge of the thousand year day of the Gods was had also by others; see Ps. 90:4; 2 Pet. 3:8. The Lord's reckoning of time is, indeed, awesome. Astronomers know of no star so massive as to require a thousand earth years to complete one rotation on its axis. Little wonder that Kolob is "First in government [and] the last pertaining to the measurement of time." (See figure 1, facsimile 2, Book of Abraham.) The sum of all the systems of time governing all heavenly bodies equals one celestial year: "And they give light to each other in their times and in their seasons, in their minutes, in their hours, in their days, in their weeks, in their months, in their years–all these are one year with God, but not with man" (D&C 88:44).
76. Beginning the day in the evening and ending it in the morning, or daylight hours, reflects the Hebraic background of the text. (The Jewish religious day is calculated from sundown to sundown.) Abraham and Moses consistently follow this formula.
77. Though one may work six days, the actual amount of time involved is closer to forty-eight than one hundred forty-four hours.
78. Scripture states that Adam lived 930 years (see Gen. 5:5; Moses 6:12). However, one manuscript of Joseph Smith's translation of Genesis stated that Adam was almost one thousand years old at his passing. This figure is supported by Edward Stevenson, who wrote that he had learned from Joseph Smith that "Father Adam began his work and finsihed what was to be done in his time, living to be one thousand years old with the exception of about six months" (Matthews, 85).
79. Joseph Fielding Smith noted: "When this earth was created, it was not according to our present time, but it was created according to Kolob's time, for the Lord has said it was created on celestial time which is Kolob's time. Then he revealed to Abraham that Adam was subject to Kolob's time before his transgression" (*Doctrines of Salvation*, 1:79; emphasis in original.)
80. Note that the reference to being obeyed is in the context of living things not yet on the earth.
81. This point is made clear in Abr. 4:29-30.
82. See chart, Sequence of Creation.

83 The Hebrew term for "brood" means to overshadow or envelop. The Spirit of the Lord enveloped the earth so as to bring forth life much as a mother hen hatches her chicks with the warmth of her own body.

84 Moses 2:2 reads that God "caused" darkness to envelop the deep after the planet was formed. This led Orson Pratt to surmise that this was why creation began in the evening. He suggested that God produced the darkness by obscuring the light of the "self-luminous" matter composing the planet. Why God would choose to do such a thing is not explained (see JD, 20:73-74). Then too, both Moses 2:3 and Abraham 4:3 simply refer to the presence of darkness at the time of the planet's initial formation without suggesting any divine causation. On balance, it seems more likely that Earth was organized in darkness (a metaphor for chaos?) and thereafter brought into light, rather than being organized in light, then hidden in darkness, only to be restored to light again.

85 See Moses 2:6-9, 20 and Abr. 4:6-9, 20. The word "firmament" (from the Latin firmamentum) is an unfortunate translation of the Hebrew raqia which suggests something extended or stretched out—an "expanse."

86 See Moses 2:14-19; Abr. 4:14-19.

87 See JD, 15:264; 16:314; 18:316.

88 The extent of Earth's first land mass is unknown. Additional land emerged subsequently; in the days of Enoch: "There also came up a land out of the depth of the sea" (Moses 7:14). Orson Pratt believed that the "submerged earth" may have existed "in the form of partial or imperfect organizations" for thousands or even millions of years before it emerged from the primordial sea on the third day of creation (see JD, 18:316).

89 See 2:160; 8:243. Moses 3:8 may mean that the seeds for the Garden of Eden were planted on the seventh day.

90 See Abr. 4:11-12, 20-25; Moses 2:11-12, 20-25.

91 It is interesting to note that the term "life" is first used by both Abraham and Moses in connection with marine life. This suggests that the higher definition of life does not apply to vegetation. See Abr. 4:20; Moses 2:20.

92 Orson Pratt commented: "On the seventh day the Bible says that God ended his work. He did not altogether [totally] end it on the sixth, but he ended it on the seventh" (JD, 15:243).

93 At this point, Abr. 5:4 and Moses 3:4 employ "day" as a collective noun for the combined period of time involved in both the spiritual and temporal creation of "the generations of the heavens and of the earth." Thus this "day" was much longer than the six days or times considered in the temporal or physical phase of creation alone. Time and eternity were combined as one creative "day."

94 Contradicting the view of geologists that animals were the first flesh on Earth, Orson Pratt declared, "But the Lord gives us different information from this. He shows us that among all the animated creatures of flesh, man was the first that was ever placed upon the earth in this temporal condition, contradicting the theories of geologists—that is, so far as placing man on the earth in this present probation is concerned." However, Pratt was uncertain as to circumstances "millions of ages before the world was organized temporally" (JD, 21:201).

95 In Abraham's account, reference to animal life follows the coming of Adam and Eve; in Moses' account, it is placed after Adam and before Eve.

96 The view that animal life preceded human life is maintained by defining "first flesh" as meaning the mortal flesh of Adam following his transgression. The context of "first flesh" in Moses's account of the creation, not of the Fall, simply will not support such tortured exegesis. Flesh can be either mortal, translated, or immortal. Appearing as a resurrected being, Jesus told his disciples, "Behold my hands and my feet, that it is I myself: handle me, and see; for a spirit hath not flesh and bones, as ye see me have" (Luke 24:39). The Father and the Son have bodies of resurrected flesh (see D&C 130:2). Both Adam and Job testified that they would behold God in their immortal flesh; see Moses 5:10; Job 19:25-27; TPJS, 326, 367.

97 *Doctrines of Salvation*, 1:77; emphasis in the original. See *Man, His Origin and Destiny*, 362-64.

98 Orson Pratt was of the same understanding: "Though it was the seventh day, no flesh but this one tabernacle [Adam] was yet formed" (JD, 21:200; see 15:243; 16:317). Among those who taught that man was on Earth prior to any animal life was John Taylor; see JD 18:327.

99 Joseph F. Smith, John R. Winder, Anthon H. Lund, "The Origin of Man," *The Improvement Era*, 13 (Nov. 1909) 78, 80.

100 *Deseret Evening News*, 27 Dec. 1913, as found in Andrus, *God, Man and the Universe*, 353.

101 See Matt. 4:4; 6:11; Luke 11:3; John 6:35, 47-41.

102 See JD, 15:262-63; 16:325.

103 See D&C 77:6. Joseph Smith said: "The world has had a fair trial for six thousand years; the Lord will try the seventh thousand Himself" (TPJS, 252).

104 Scriptural literalists reason that when the seven thousand years of the creation week are added to the six thousand years of human history since the Fall, the overall age of the organized physical planet is approximately thirteen thousand years. Estimates of time since the Fall are in exact (see JD, 2:235; 14:351; 15:262; 16:324; 19:286; 21:201; 25:61; 26:200; Andrus, *God, Man and the Universe*, 315-16, 329; and *Science and Mormonism*, ch. 8).

105 See The Articles of Faith 10; D&C 88:41; Moses 1:5. The physical-spiritual character of the pre-Fall earth is the basis for it being described both in paradisiacal and terrestrial terms. It is sufficient to say that it was a spiritual and, therefore, an immortal world.

Chapter Six

The Fall of Earth

In the beginning, Earth was a spiritual-physical sphere. Slowly rotating on its axis once every thousand years, it appeared almost motionless in the enlivening light of Kolob which bathed it in an enveloping paradisiacal glory. It was the mountain of the Lord; he walked upon it.[1]

Edenic Conditions

Earth's capitol was the Garden of Eden, but its atmosphere prevailed everywhere. All creation thrived under the benevolent reign of Adam, its heaven-anointed king. Like a father, he gave names to the denizens of the animal kingdom. And like gentle children, they accepted his kindly rule over them.[2] Neither was a threat to the other; life was the natural order; of things. Discord, disease, and suffering were yet to make their appearance and death was still a stranger. Orson Pratt said:

> Every particle of air, of water, and of earth, was so organized as to be capable of diffusing life and immortality through all the varied species of animated existence—immortality reigned in every department of creation; hence it was pronounced very 'good' (JD, 2:281).[3]

Parley P. Pratt, mindful of Isaiah's description of future millennial conditions, described this primeval paradise as a world where:

> . . . everything that grew was just calculated for the food of man, beast, fowl, and creeping thing: and their food was all vegetable. Flesh and blood were never sacrificed, to glut their souls or gratify their appetites; the beasts of the earth were all in perfect harmony with each other; the lion ate straw like the ox, the wolf dwelt with the lamb. . .all was peace and harmony, and nothing to hurt nor disturb in all the holy mountains (*VW*, 86).

The harmony between man and creature was reflected in its vegetation. It was a weedless world. The many troublesome and harmful plants so common today—plants reflective of a fallen, telestial sphere—were not to be found. Producing in her strength, the earth affirmed the goodness of God and his pleasure in those things which strengthen the body, enliven the soul, please the eye, and gladden the heart.[4]

Nor was the immortal planet subject to the extreme climatic variations which characterize this present age. Instead of zones of frigid cold and torrid heat, a stable, balanced climate prevailed from pole to pole. Such conditions precluded any violent meteorological activity such as hurricanes, tornadoes or severe electrical storms. In all likelihood, there were only infant winds—light breezes.[5] Earth's vast coal and oil deposits, even beyond the arctic circle, are indicative of ancient luxuriant forests, teeming with a myriad variety of plants and animals. It was a divine age; tranquility was the hallmark of every dimension of creation. Only with the passage of time did that splendid epoch give way to savage conditions. Our present harsh world is a product of devolution over a period of many centuries, a devolution marked by increasing violence and catastrophe as Eden faded from view.

Earth, like the Sabbath, was made for man, not man for Earth. It exists because man exists. It is subject to a higher authority than impersonal natural law, being the obedient servant of the Priesthood. Adam was its God-appointed surrogate. He received the keys of the Holy Priesthood—including "the keys of the universe"—before this world was organized.⁶ As lord of creation, Adam exercised universal dominion and stewardship. The destiny of Earth and every living thing upon it was bound up in him: Adam's course was to be the course of all.

The Fall of Man

According to both Abraham and Moses, Adam was alone in the Garden of Eden when he received the first commandment of record:

> Of every tree of the garden thou mayest freely eat. But of the tree of the knowledge of good and evil, thou shalt not eat of it, nevertheless, thou mayest choose for thyself, for it is given unto thee; but, remember that I forbid it, for in the day thou eatest thereof thou shalt surely die (Moses 3:16-17; see Abr. 5:12-13).⁷

Thereafter, Adam was provided with a companion, "an helpmeet," whom he characterized as "bone of my bones, and flesh of my flesh," not only because she, unlike the animals, was of the *Man*kind, but because she was his eternal wife. In time, and according to the foreordained plan of salvation, Earth's first and only immortal couple transgressed the commandment, thereby bringing upon themselves and their posterity the judgment of God—the natural consequences of violated law.⁸

The spirit substance which insured their immortality was purged from their systems when they partook of the incompatible, blood-producing fruit. Over a period of time their body chemistries were altered as they literally ate themselves into mortality.⁹ A natural law had been broken; a natural consequence ensued. The fact that Adam could have purged blood (mortality) from his system and regained his lost immortality by partaking of the fruit of the tree of life indicates the literalness, the reality, of the Fall.¹⁰ So it was "paradise lost" for Adam and Eve.¹¹ The fruits of the Garden which had sustained the once immortal couple were no longer available to them. The tree of life had become the "forbidden fruit."¹²

A Literal Fall

The transgression of natural law does not, of itself, produce spiritual death;¹³ only disobedience to the commandments can do that. Consequently, the Lord not only warned Adam against the lethal effects of the toxic forbidden fruit, he also made it a point to forbid its use, thereby introducing moral law and sin into the situation.¹⁴ Had he *not* done so, Adam, in becoming mortal, would have remained spiritually alive. But this would have resulted in an unacceptable, partial Fall and, therefore, in an unacceptable partial redemption. The Father's work is *both* the immortality and the eternal life of man. Had Adam given his posterity only mortal life, all would have ended in death and been for naught. By joining them in spiritual death he pointed the way to life and salvation through Christ by his own repentance and spiritual rebirth.¹⁵ Therefore, it was foreordained that Adam should fall both physically and spiritually so that the Redeemer, through his atoning sacrifice and resurrection might assure mankind of *both* immortality and eternal life. The fall of Adam and the atonement of Jesus were the two sides of one coin: the perfect plan of salvation.

The Atonement is the direct effect of the Fall. To doubt the one is to doubt the other. They are bi-conditional realities; they stand or fall together. Indeed, the genius of the gospel is that it is a series of bi-conditional propositions. Each depends upon the others for its validity. Paul wrote of the "man" Adam and the "man" Jesus: "For since *by man* came death, *by man* came also the resurrection of the dead. For as *in Adam* all die, even so *in Christ* shall all be made alive" (1 Cor. 15:21-22). If mankind did not die because of Adam's transgression, how can it live because of Christ's atonement? If there was no *cause,* how could there be an *effect?*

The Garden of Eden

While the entire pre-Fall earth was enveloped in paradisiacal glory, the Garden of Eden was especially blessed in having God commune there with Adam and Eve.[16] The Garden symbolizes the relationship all mankind once enjoyed with God and which will be had again by as many as qualify for it through the merits of the Savior. Its beauty and perfection also reflected the spiritual natures of its two innocent inhabitants. Virtue garnished their thought's unceasingly; their hearts and minds were pure and uncorrupted. Eliza R. Snow wrote:

> So near allied to beings o'er the sky,
>   Their minds were holy—every thought was high:
> The stream of knowledge then was pure and broad,
> For man held converse with th' Eternal God.[17]

The Garden of Eden was located in the region encompassing what is now Independence, Jackson County, Missouri—making America the oldest inhabited land on earth. This incredible fact was revealed through Joseph Smith in connection with the designation of Independence as the site of the future city of Zion, the New Jerusalem.[18] Brigham Young remarked:

> Now it is a pleasant thing to think of and to know where the Garden of Eden was. Did you ever think of it? I do not think many do, for in Jackson County was the Garden of Eden. Joseph has declared this, and I am as much bound to believe that as to believe that Joseph Smith was a prophet of God.[19]

America is the choicest of all the lands of promise because it is the site of the Garden of Eden—the holiest spot on earth.[20] For this reason alone Zion, the New Jerusalem, cannot be built anywhere else; it must and will be located in the area of Independence, Missouri prior to the second advent of the Savior.[21] In the beginning, God walked and talked with his children in the Garden; he will do so again as part of the restoration of all things. After being driven from the Garden, the exiled pair settled some seventy miles north of Independence in what is now Daviess County, Missouri. In man's pristine language their new home was called Adam-ondi-Ahman.[22]

Mankind's departure from the literal presence of the Lord is symbolized by the expulsion of Adam and Eve from the Garden. They were removed from the Garden and, in time, the Garden was removed from this planet.[23] Its site now lies buried beneath the asphalt and concrete of a suburb of Kansas City where it is trodden by the unhallowed feet of men and desecrated by the banal sprawl of twenty-first century civilization.[24]

The Natural Effects of the Fall

When Adam, the lord of creation, fell his entire stewardship fell as well. Only the spirit earth remained untouched. But what had been spiritual and immortal on this planet became temporal and subject to the physical laws which govern temporal organizations.

Fall from Celestial Time

Celestial time ended and telestial time began when Adam and Eve fell from the presence of the Lord and were sent forth as exiles from the Garden of Eden, from eternity into time, from paradise into the lone and dreary world. Not until then did

Adam receive his own reckoning of time. No longer did Earth move in harmony with the rhythm of Kolob. The speed at which it rotated on its axis was increased by a factor of 365,000! Its thousand year days were over, being foreshortened to a mere twenty-four hours.[25] This acceleration of time, this speeding up of Earth's heartbeat, was indicative of its fall from immortality. Earth's longevity had become finite; its days were literally numbered. Instead of living forever, its life span was calculated in terms of one celestial week. Thus Earth fell from the time measurement of the highest to that of the lowest order of worlds. Indeed, even the moon is superior to it in this respect (see Abr. 3:5).

Fall from Primeval Light

No longer did Earth bask in the glorious spiritual light of eternal worlds. In being cast down to this solar system it, like man, was brought into the presence of the gross, physical light of the sun, moon, and stars.[26] The sun, whose organization is associated with the fourth day of creation, was to became the source of both light and life for the earth throughout its temporal existence. Brigham Young stated: "When the Lord said—'Let there be light,' there was light, for the earth was brought near the sun that it might reflect upon it so as to give us light by day, and the moon to give us light by night" (JD, 17:143). Although Genesis 1:3 actually applies to Earth's first source of light, in this instance, Young simply broadens the passage to apply to the now prevailing second source of light to which the planet was "brought near" after the Fall.

Fall from Primeval Place

> Thou, Earth wast once a glorious sphere
>   Of noble magnitude,
> And didst with majesty appear
>   Among the worlds of God.[27]

Brigham Young agreed with these sentiments of Eliza R. Snow:

> When the earth was framed and brought into existence and man was placed upon it, it was near the throne of our Father in heaven. . . .but when man fell, the earth fell into space, and took up its abode in this planetary system, and the sun became our light (JD, 17:143).[28]

Man is the key to the universe; his destiny dictates the destiny of all things. Therefore, Adam's spiritual and temporal fall imposed a similar fate upon the heavens, the earth and all of their hosts. As he was driven from the Garden of Eden, from the presence of the Lord, so was this planet driven from it primeval setting in God's presence.[29] If Kolob and the "residence of God" are to be found somewhere in the dense nucleus of stars located in the hub of the Galaxy then, in falling, this planet was transported a distance of approximately *thirty thousand light years* out to this insignificant solar system located in the suburbs of the Milky Way.[30]

And when Earth fell, so did its counterpart heaven. Since the original firmament or atmosphere enveloping the planet was compatible with the bloodless, immortal natures of man and beast, it is likely that the "air" they breathed was more refined than the blood-transmitted oxygen which is the basis of mortal life. It may have been a life-sustaining element akin to, if not identical with, the Spirit of the Lord.[31] The placing of Earth in this solar system led to the necessary modification of its atmosphere so as to make it compatible with the oxygen-breathing requirements of the earth's then-mortal denizens.[32] In any event, both Earth and man are cut off from the presence of the Lord by a very real veil of element (see *VW*, 87). The Savior's return will be attended by an unveiling of those glories which now lie hidden behind a curtain of gross materiality.

Fall From Immortality

The existence of plant and animal fossils which are thought to have originated in the so-called paleozoic era of geologic time[33] would appear to contradict the scriptural declaration that death was unknown on this earth until after the Fall.[34] In an effort to reconcile Scripture with paleontology it has been suggested that plant and animal life antedated the coming of Adam by hundreds of millions of years in which the natural processes of birth and death were taking place on the pre-Adamic Earth. Then God placed man in a special sector of the planet—the Garden of Eden—where the natural order of things was suspended during the pre-Fall period. In other words, God created an oasis of immortal life in a global graveyard of death.

Such an Hegelian-like synthesis is appealing because it allows the scientific thesis of a mortal world to be joined to a scriptural anti-thesis of an immortal creation. Unfortunately, the synthesis—as is usually the case where revelation and reason clash—is achieved at the expense of revelation. As we have seen, Earth was organized in a totally different sphere than that in which it now finds itself. It was both a physical and a *spiritual* body. Having been organized by the Lord, it partook of his own excellence: "And I, God, saw *everything* that I had made, and behold, *all things* which I had made were *very good*" —meaning God-like: perfect and immortal.

One might as well argue that a mortal body possesses an immortal heart as to suggest that death reigned planet-wide except in a small paradise in the land of Eden. Joseph Smith explained that the book seen by John (Rev. 5:1) symbolized the things of God "concerning this earth during the seven thousand years of its continuance, or its *temporal existence*" (D&C 77:6). This statement admits of no other interpretation than that the *entire planet,* not the Garden alone, became a temporal organization approximately seven thousand years ago.[35] Then too, the whole tenor of the creation accounts in Scripture is to the effect that the Gods were dealing with the planet as an entity.[36] Their labors pertained to all life, to "the generations of the heaven and the earth" as such, and not simply to a select group of immortals.

Nor was Adam's dominion limited to his stewardship over the animals found in the Garden, it was "over the fishes of the sea, and over the fowl of the air, and over the cattle, and over *all* the earth" (Moses 2:26; see Abr. 4:26). There could have been no universal fall if Adam had lacked universal dominion or if the major portion of the planet and its creatures were in a mortal condition long before Adam himself fell to that state.[37] On the issue of death, no honest synthesis of prevailing scientific thought with the clear declarations of Scripture is possible.

The Fall of Nature

In the beginning, Earth's vegetation was wholly for nourishment and beauty: "Out of the ground made the Gods to grow every tree that is pleasant to the sight and good for food" (Moses 2:26; see Abr. 4:26). But the Fall deprived Earth of the spiritual nature which rendered it immortal; it was no longer sustained by the enlivening power of the Spirit of God to the same high degree it formerly enjoyed. Instead, it had become a mortal world, a fallen, telestial kingdom ruled by a set of lesser laws appropriate to its new condition (see D&C 88:26-39). As such, it was diminished in its life-giving powers so that the fruits of the Garden which formerly sustained Adam and Eve were no longer available to them. Spontaneity shifted from the production of fruits, flowers, and desirable vegetation to the growth of troublesome weeds and poisonous plants characteristic of Earth's present condition: "Thorns also, and thistles shall it bring forth to thee, and thou shalt eat the herb of the field" (Moses 4:24).[38] The tree of life was now the forbidden fruit; the tree of knowledge was everywhere!

The introduction of alien telestial vegetation suggests another possibility: the introduction of alien animal life as well. While it is assumed that all creatures are descendants from pre-Fall species, such may not be the case. Certain animals may have been as out of place on a paradisaic world as the thorns and thistles. For example, the dinosaur could well be a latecomer to this planet, arriving after the Fall only to be destroyed in the Flood. There are degrees of glory among animals as well as men. Since the different species are assigned to appropriate kingdoms after the resurrection, why not before as well?[39] In any event there has been a profound descending for the many "kingdoms" on this planet. Like Adam, they possess life, yet they experience death. Earth has become a heterogeneous realm of oppositions, variations, and contradictions.

The substitution of blood for spirit profoundly affected the dispositions of living things. As time passed, some of the formerly harmless companions of man became threatening enemies. Creatures that had once lived side by side with their human keepers in mutual security either fled at the sound of their approaching step or poised for attack. The lower order of vegetation, the "green herb," had been organized to sustain the higher kingdoms of man and animals:

> And I, God, said unto man; Behold, I have given you every herb bearing seed, which is *upon the face of all the earth,* and every tree in the which shall be the fruit of a tree yielding seed; to you it shall be for meat. And to every beast

of the earth, and to every fowl of the air, and to every thing that creepeth upon the earth, *wherein there is life,* I have given every green herb for meat: and it was so (Gen. 1:29-30; see JD, 16:358).

Thus, like man, the entire animal kingdom was herbivorous in the beginning. Neither purchased life at the price of another's death. The Fall comprised this harmonious arrangement. The wily serpent which had willingly served as the instrument of Satan's first attack on mankind, had, in doing so, become the "Judas goat" of all animal creation as well—peaceful munching and chewing gave way to clawing, ripping, and tearing. The bounds of the several kingdoms were breached; the divinely established order was disrupted.[40] Kingdom rose up against kingdom, species against species. The spoilers had arrived; man invaded and was invaded in turn. The Spirit of the Lord, of peace, no longer prevailed. World war had begun.[41]

SUBSEQUENT CURSES AND FALLS

Violation of the laws of God produces consequences which almost always affect the innocent as well as the guilty—the sins of the fathers are visited upon the children even as the sins of the children are often visited upon their fathers.[42] While a sin may be isolated, its bitter fruits seldom are; sooner or later, one way or another, others are obliged to partake of them. Earth was no exception. Mother Earth and her Redeemer are alike in that, while both were innocent, both were willing to be cursed for the sake of others. The curse of the Fall was visited upon the earth because of Adam's transgression, even as the curse of the Atonement was visited upon the Redeemer because of the sins of mankind.[43] But whereas the Redeemer descended below all things in one final, climactic sacrifice, the earth has been subjected to additional curses following her expulsion from the presence of God. Man's disobedience has swept him over one moral and spiritual cataract after another. The Fall of man has been followed by the falls of men; the earth has been cursed over and over again because of them.

Cain

Earth was first defiled by Cain's murder of his brother Abel. So heinous a crime had never before been committed among the children of Adam and Eve.[44] Murder! The blood of innocence had been wantonly shed in what is probably the most outrageous act of rebellion against God in human history.[45] Earth, the source of the physical body of him whose blood now seeped into her flesh, turned against Cain the husbandman whose livelihood she had made possible.

> And the Lord said: What hast thou done? The voice of thy brother's blood cries unto me from the ground.[46] And now thou shalt be cursed from the earth which hath opened her mouth to receive thy brother's blood from thy hand. When thou tillest the ground it shall not henceforth yield unto thee her strength. A fugitive and a vagabond shalt thou be in the earth (Moses 5:35-37).

Cain was forced to wander the alienated planet which would no longer respond to his labors. He fled the land God had cursed against him and built a city named for his son, Enoch—the first city known to man.[47]

The Lord's judgment was altogether just. Cain, like his mentor, Satan, was a liar and a murderer from the beginning.[48] He was the original "Gadianton," being the founder of the organized kingdom of the devil among men.[49] The murder of Abel unleashed a flood of conspiratorial evil upon the human family that has brought misery and death to uncounted millions of innocent souls swept away in its wake. The family of Cain provided the nucleus around which Satan marshalled his offensive against the kingdom of God. It is a tale of blood and horror. Cain's grandson, Irad, was murdered by his great grandson Lamech for violating his oath of secrecy as a member of Cain's conspiracy (see Moses 5:49-50). Betrayed by his own wives, Lamech, like Cain, became an outcast. Still the forces of evil prospered; the kingdom of the devil spread its contagion throughout the world:

> And thus the works of darkness began to prevail among all the sons of men. And God *cursed the earth with a sore curse,* and was angry with the wicked, with all the sons of men whom he had made; For they would not hearken unto his voice, nor believe on his Only Begotten Son, even him whom he declared should come in the meridian of time, who was prepared from before the foundation of the world (Moses 5:55-57).

The People of Canaan

The "sore curse" God imposed on Earth in Lamech's generation (probably in the late third century after the Fall) was followed by other instances of divine judgment. Enoch prophesied that the "people of Canaan" (descendants of Cain) would launch a war of aggression against the weaker "people of Shum" and utterly destroy them. However, the Lord would turn the wine of victory to gall by cursing the conquered and occupied land:

> The land shall be barren and unfruitful, and none other people shall dwell there but the people of Canaan; for behold, the Lord shall *curse the land with much heat,* and the barrenness thereof shall go forth forever; and there was a blackness came upon all the children of Canaan, that they were despised among all people (Moses 7:7-8).

Enoch

Enoch was instructed to carry the message of salvation to many lands with the Lord's warning: "Repent, lest I come out and smite them with a curse and they die" (Moses 7:10) The faith of Enoch was such that the planet was responsive to his every command. With the spoken word alone he caused the earth to tremble, the mountains to flee, and the rivers to alter their courses. "All nations feared greatly, so powerful was the word of Enoch, and so great was the power of the language which God had given him."[50] This mighty patriarch, like many prophets after him, was also a military leader who led the people of God against their enemies, "and there went forth a curse upon all people that fought against God" (Moses 7:15). And so judgment followed judgment in a mounting crescendo generation after generation.

Methuselah

Another "sore curse" was visited upon the planet in the days of Methuselah:[51]

> And there came forth a great famine into the land, and the Lord cursed the earth with a sore curse, and *many* of the inhabitants thereof died (Moses 8:4).

Whether this particular judgment triggered a general decline in the earth's fertility is unclear. However, such a decline did occur. In blessing his son Noah, Lamech prophesied: "This son shall comfort us concerning our work and toil of our hands, because of the *ground which the Lord hath cursed"* (Moses 8:9). Saint as well as sinner eventually suffered under the Lord's judgments, so that the planet, which was initially cursed against one man was finally cursed against them all.

Sodom and Gomorrah

We would not know that a series of curses or judgments were visited upon the earth prior to the Flood were it not for the revelations given to the Prophet Joseph Smith in the writings of Moses. Genesis mentions only the incident of Cain. However, Genesis does record the occasion after the Flood when God cursed the notorious cities of Sodom and Gomorrah for their abominations. Abraham pled in vain for the preservation of Sodom when not even ten righteous men could be found in the entire city.[52] The degeneracy of the men of Sodom was such that they attempted to assault the two angels sent to save Lot and his family.[53] When the sun arose the next morning, Lot and his two daughters (Lot's wife and sons-in-law had perished) were entering the village of Zoar.

> Then the Lord rained upon Sodom and upon Gomorrah brimstone and fire from the Lord out of heaven; And he overthrew those cities, and all the plain, and all the inhabitants of the cities, and that which grew upon the ground. . . .And Abraham got up early in the morning to the place where he stood before the Lord: And he looked toward Sodom and Gomorrah, and toward all the land of the plain, and beheld, and, lo, the smoke of the country went up as the smoke of a furnace (Gen.19:24-28).

Today, the fertile valley of Siddim, where once stood Sodom and Gomorrah, lies buried beneath the acrid waters of the Dead Sea. All around lies barren desert, rocky desolation, and the silence of an empty land—mute, but profound testimony to the fact that the wages of sin is always death.

Famine

The foregoing incidents combine to show one thing: nature turns against man when man turn against nature's God. The "natural man," who is an enemy to God, is eventually cursed pertaining to the good things of the earth upon which he places so much stress and for which he commits so many crimes. Note that God's "curses" do not take the form of the exotic spells or the mysterious happenings found in fairy tales. Rather, they strike at man's physical life and his ability to sustain it. This is why God's curses usually involve natural calamities.

Famine-inducing drought is the natural disaster most frequently mentioned in Scripture.[54] Indeed, the Lord has employed it to bring about the desired migration of men and nations. A prophesied famine befell the Chaldeans in the days of Abraham prompting his family's move to Haran in Mesopotamia.[55] It was still raging when Abraham left Haran and journeyed to Canaan: "And there was a continuation of a famine in the land; and I, Abraham concluded to go down into Egypt, to sojourn there, for the famine became very grievous" (Abr. 2:21). A famine in the days of Isaac led him to consider migrating to Egypt until the Lord instructed him to remain in Canaan with the assurance that if he did so, the blessings of Abraham would become his blessings as well.[56] Had Isaac disobeyed the Lord, the story of Israel would have been quite different: Israel would have been founded in Egypt rather than in the land of promise. This, the Lord did not intend, for Israel was to be established in accordance with the covenants Jehovah had made with Abraham.

So, at the right time, famine brought Jacob to Egypt as it had his grandfather Abraham more than a century earlier. The house of Israel followed Joseph into a strange land in search of survival—not merely, as they supposed, the survival of their lives, but more importantly their survival as a covenant people. For had Israel remained in Canaan, evidence suggests that they would have been either destroyed by their enemies or absorbed into the all-too-enticing culture surrounding them.[57] But as sojourners in Egypt, they were safe from the Canaanites and were permitted to live apart from the general society, thereby giving them time to gain strength and to establish their identity as a unique nation. The famine that made Joseph governor of Egypt and his family subject to his righteous stewardship was a blessing in disguise.

Famine is a frequent motif in the story of ancient Israel. It accounts for Ruth (a Midianite) marrying an Israelite.[58] It was inflicted on Israel for Saul's slaughter of the Gibeonites.[59] Elijah caused a famine when he sealed the heavens so that there was no rain for three and one half years.[60] Other references make it clear that God has cursed the earth with famine in the past and will do so in the future.[61]

A Testimony to Sin

Parley P. Pratt wrote:

> The great curses which have fallen upon the different portions, because of the wickedness of men, will account for the stagnant swamps, the sunken lakes, the dead seas, and great deserts (*VW*, 87).

It is apparent that at least some of this planet's bleak, inhospitable regions were once the dwelling places of very sinful men and women. Their defilement of themselves brought curses upon the lands they occupied. And as they moved from place to place, seizing the most desirable areas for themselves, they spread their moral contagion. This caused the curses of God to follow them, thereby intensifying the fall of the earth from her primeval glory.

Man's offenses against God compounded man's offenses against himself. For as more and more of the planet became less productive due to divine judgments, man turned with increasing fury and frequency against his fellow man in competition for earth's diminishing spoils.[62] Contrasting the ways of men with the ways of God, Joseph Smith wrote of the desolation left by advancing armies down through time:

> The greatest acts of the mighty men have been to depopulate nations and to overthrow kingdoms; and whilst they have exalted themselves and become glorious, it has been at the expense of the lives of the innocent, the blood of the oppressed, the moans of the widow, and the tears of the orphan. . . .before them the earth was a paradise, and behind them a desolate wilderness; their kingdoms were founded in carnage and bloodshed, and sustained by oppression, tyranny, and despotism (TPJS, 248).

It is noteworthy that history's first wars began immediately after the initial cursing of the earth in the days of wicked Lamech about three hundred years after the Fall.[63] Fallen man had finally come of age. Satan raged in the hearts of men and men raged in their hearts against one another.[64] James asked, "From whence come wars and fighting among you? come they not hence, even of your lusts that war in your members?" (James 4:1). Would the wicked, in being cursed, remain on the least desirable portions of the earth if it were in their power to appropriate the uncursed lands of others? Is this not a prime cause of war? And this will remain a fact of human life as long as wicked men rule and the earth remains the cursed planet that it is.

When it came forth from the hand of its Maker, the paradisiacal Earth was an unflawed work of perfection. There were no steaming jungles, no frozen wastelands, no barren deserts, no towering mountains burdened with perpetual snow and ice. Such post-fall conditions came about in a series of changes over many centuries. But now Earth is a living testimony to the sins of man and the judgments of God.

Our Aging Earth

Since God lives in absolute harmony with truth and law, all that he does in nature is positive. Nothing negative can be attributed to him as an arbitrary action on his part. Consequently, mortality, with its many undesirable aspects, is a product of broken law. Transgression produced the Fall and the Fall produced mortality.[65] This applies to the earth today: the burdens of mortality have taken their toll on our ageing planet. Mother Earth has experienced many of the symptoms of change and decay characteristic of us humans. The fevers of forest fires and burning deserts, the thirst of drought, the barrenness of lifeless lands, the vomiting of volcanic eruptions, the sweating of floods and tidal waves, the baldness of denuded woodlands, the convulsions of earthquakes, the organic distresses of inward groanings, the weeping of rains, and even the amputation and separation of portions of her body—all these have been her lot. Little wonder Enoch heard her cry out, "Wo, wo is me."[66]

While her fall was unavoidable, mankind has compounded it with his many reckless assaults on her treasures. She has been plundered by some men and nations with the same thoughtless, unconscionable disregard that characterized the tomb robbers of Egypt. Nothing was sacred, nothing mattered but the treasure itself. The ignorant despoilers' only thought was the wealth, ease, gratification, and power the tombs could provide them. In their anxiety to seize the things they esteemed to be of greatest material value, they trampled, smashed, scattered, or carried off as common utensils and household wares, objects of priceless value.

Modern technology has made a science of the exploitation of this planet. It has destroyed vast regions of virtually irreplaceable forest and their native species, leaving the land scarred and disfigured after extracting coal, iron, copper, and other valued minerals. It alters and unbalances her very being by ceaselessly pumping oil and natural gas from deep within her bowels. It clogs her veins and arteries—once pure, sparkling oceans, rivers, lakes and streams—with all manner of deadly chemical pollutants. It has contaminated the atmosphere, filling it with the acrid, poisonous haze of commerce and industry. It has brought much of the life which once teemed in her waters and flourished on her lands and in her forests to the brink of extinction. Man has even become a cosmic litterbug, littering the heavens with thousands of bits and pieces of space debris. While it is true that, to a degree, such activities are the price we must unavoidably pay for civilized living, yet we must admit that much that has been done in the name of progress was and is done in an uncivilized manner.

It is argued that we are justified in utilizing the earth's waters, forests, lands, minerals, and animal life as we sees fit. For how can we subdue the earth without using the earth? But the issue is not our right to use, but our *right use* of the planet. We should exercise dominion over the earth not as owners, but as stewards:

> For it is expedient that I, the Lord, should make every man accountable, as a steward over earthly blessings, which I have made and prepared for my creatures. I, the Lord, built the earth, my very handiwork; and all things therein are mine.[67]

Earth does not belong to us; we are tenants on another Man's property, stewards over another Man's wealth. As such, we will be obliged to give an accounting of our stewardship over everything placed under our care. Earth was prepared as a blessing for all of its inhabitants. The Father never designed that the strong should possess it at the expense of the weak. Had his children loved their brothers and sisters as themselves and regarded the earth with gratitude and respect, misery and want would have been virtually unknown to mankind. But because men have sinned grievously against God, the planet is cursed, and because the planet is cursed, the curses of war, famine, disease, and pestilence continue to stalk the land. We have reaped what we have sown. And the end is not yet.

NOTES

1. He will walk upon it again when it regains its lost glory and fulfills the prophecy of Isaiah; (see Isa. 11:6-9; 65:25; 2 Ne. 2:22).
2. See Turner, *Woman and the Priesthood,* Chapter III.
3. Pratt explained that the earth was pronounced very good because "it contained every necessary ingredient to render happiness to the beings who were designed to occupy it" (JD, 1:328).
4. See D&C 59:18-19; *VW,* 86.
5. The first reference to wind in the Bible is made in connection with the Flood; see Gen. 8:1.
6. See Abr. 4:26-28; Moses 2:26-28. Joseph Smith taught: "The Priesthood was first given to Adam; he obtained the First Presidency, and held the keys of it from generation to generation. He obtained it in the Creation, before the world was formed, as in Genesis 1:26, 27, 28. He had dominion given him over every living creature. He is Michael the Archangel, spoken of in the Scriptures" (TPJS, 157).
7. Both accounts are silent as to when Eve learned of this commandment.
8. Known as the "first judgment." See 2 Ne. 9:6-7 and Turner, *Woman and the Priesthood,* 42-43.
9. "When Adam and Eve had eaten of the forbidden fruit, their bodies became mortal *from its effects,* and therefore their offspring were mortal" (Brigham Young. JD, 1:50; emphasis in original;. see 1:282; 10:235).
10. See Alma 12:26; 42:3-5.
11. How long Adam and Eve remained in the Garden of Eden before transgressing the second commandment is unrevealed.
12. In all that they did, Adam and Eve were acting for and in behalf of all mankind, doing symbolically the things fallen men and women would do. Thus, they transgressed both spiritual and physical law because mankind would do so. The key to their actions is found in the infinite atonement of Jesus Christ for the sins of the world.
13. Spiritual death is alienation from God. It ranges from banishment from his literal presence to the total loss of the influence of his Spirit. Adam suffered the former; Satan will suffer the latter. See Alma 42:7-9; D&C 29:36-42; 76:36-37; Rev. 20:9-10.

14 Orson Pratt spoke to this point; see JD, 1:282-83.
15 See Moses 6:51–68.
16 By the same token, the grove where Joseph Smith beheld the Father and the Son was rendered sacred and more glorious by their presence.
17 Eliza R. Snow, "Eden," *Poems, Religious, Historical, and Political* (Liverpool: 1856.), 153.
18 See D&C 57:1-3.
19 *Journal History*, 15 March 1857, as quoted in John A. Widstoe, *Gospel Interpretations*, 254. See also JD, 8:67, 195, 211; *The Contributor,* 14:7-9. Heber C. Kimball taught the same: "and I will say more, the spot chosen for the Garden of Eden was Jackson County, in the state of Missouri, where Independence now stands" (JD, 10:235; see 8:195). However, Orson Pratt said, "I do not know that the Garden of Eden was here [America]" (JD18:69).
20 The Garden of Eden is the true and original hierocentric point—the holy center—where all things began. Certain world religions have borrowed the concept and designated such cities as Rome, Jerusalem, and Mecca as their holy centers.
21 See D&C 97:19; 101:17-21; TPJS, 85-86.
22 See D&C 116; 78:15-16. John Taylor heard Joseph Smith identify Adam with the Daviess County site: "Indeed, it was stated by the Prophet Joseph Smith, in our hearing, while standing on an elevated piece of ground or plateau near Adam-ondi-Ahman (Daviess County, Missouri), where there were a number of rocks piled together, that the valley before us was the valley of Adam-ondi-Ahman; or in other words, the valley where God talked with Adam, and where he gathered his righteous posterity as recorded in the above revelation [D&C 107:53-57], and that this pile of stones was an altar built by him when he offered up sacrifices, as we understand, on that occasion" (*GK*, 102). A gathering of an even more momentous nature will take place there in fulfillment of Daniel's prophecy (Dan. 7:9-10) prior to the world coming of Christ (see D&C 116; *DS*, 3:13-14).
23 According to Brigham Young, it still exists in its spiritual state. See JD 14:231.
24 Fortunately, the general area designated for the temple is used for religious purposes, being under the control of the LDS Church and certain splinter groups.
25 If the pre-Fall planet received light from Kolob the same way it now receives light from the sun, for half a celestial day, five hundred years, half the earth would have been in darkness, half in light. Adam and Eve may then have dwelt in light throughout their stay in the Garden of Eden and been sent forth as evening descended upon it.
26 The diminished glory of the earth stems in part from the fact that the

sun itself is obliged to "borrow its light from Kolob through the medium" of still another "governing power." See the explanation of figure 5, facsimile 2, in the Book of Abraham.

27 Eliza R. Snow, "Address to Earth." This poem was published in England in the *Millennial Star* 13 (1 September 1851): 272. The doctrinal acceptability of its contents was apparently recognized since it appeared in numerous editions of the LDS Hymnal for at least fifty years (1856-1905). Eliza R. Snow was the sister of Lorenzo Snow, the fourth president of the LDS Church, the plural wife of Joseph Smith and, subsequently, of Brigham Young.

28 Charles L. Walker recorded hearing Brigham Young state that "when this world was first made it was in a close proximity to God. When Man sinned it was hurled Millions of Miles away from its first Position, and that was why it was called the Fall." (Journal,13 July 1862, 1:234.) John Taylor also maintained that the earth "Had fled and fallen from where it was organized near the planet Kolob" (*The Mormon*, 29 August 1857).

29 Brigham Young "gave it as his opinion that the Earth did not dwell in the sphere in which it did when it was created, but that it was banished from its more glorious state or orbit or revolution for man's sake" (Record of Acts of the Quorum of the Twelve Apostles), 1849 Record Book, 41, Church Archives, as quoted in Ehat and Cook, 84; see also 60, n. 12.

30 This is a distance of 174,000 *trillion* miles!

31 President Lorenzo Snow told the saints in St. George, Utah: "The whole earth is the Lord's. The time will come when it will be translated and be filled with the spirit and power of God. The atmosphere around it will be the spirit of the Almighty. We will breathe that Spirit instead of the atmosphere we now breathe" (*MS*, 61:546; 31 Aug. 1899).

32 The indivisible character of the earth and its heaven is affirmed by the fact that "heaven and earth shall pass away," only to be followed by "new heavens and a new earth." See Matthew 24:35; D&C 29:23-24; Ether 13:9.

33 The paleozoic era extends from approximately 600 to 225 million years ago.

34 See Gen. 2:17; Moses 3:17; 6:48, 59; 2 Ne. 2:22; JD, 16:358.

35 In fact, there is no information on the fate of the Garden of Eden. It was probably removed from the earth shortly after Adam's expulsion from it. See Moses 5:2.

36 The Garden of Eden consisted of a comparatively small area of land in a larger locality called Eden. Cain was later banished to a land of exile called "Nod" which lay to the east of Eden; see Moses 5:41.

37 The Fall was as universal as was the Atonement; they were infinite acts

by two immortal beings. See 1 Cor. 15:22. Then too, if the earth did not undergo a universal fall, there would have been no need for a universal baptism of water, fire, and the Holy Ghost. These three events are inextricably bound together.
38 Brigham Young: "The thistle, the thorn, the briar, and the obnoxious weed did not appear until after the earth was cursed" (JD, 1:50; emphasis in original; see *Voice of Warning,* 87).
39 The "four beasts" of John's vision (Rev. 4:6-9) represent the glory of the classes of beings in their destined order or sphere of creation, in the enjoyment of their eternal felicity (D&C 77:3).
40 See D&C 88:36-39.
41 See Gen. 9:2-5.
42 See Ex. 10:5; 34:7; Num. 14:18; Deut. 5:9.
43 "Christ hath redeemed us from the curse of the law, being made a curse for us; for it is written: Cursed is everyone that hangeth on a tree" (Gal. 3:13; see Isa. 53; Luke 22:42-44; Rev. 19:15; D&C 19:15-19; 133:50-51).
44 Satan, not Cain, authored the idea; see Moses 5:29-31.
45 While the crucifixion of Jesus is the supreme example of the shedding of innocent blood, its perpetrators lacked Cain's knowledge of God. For this reason, Cain is Perdition while those who crucified Jesus are subjects of salvation (see Moses 5:18-25; Acts 3:13-19; TPJS, 339).
46 The earth is defiled and justice is outraged when innocent blood is shed. That blood cries out for justice and if those who are responsible for administering justice ignore that voice, they become accessories to the crime. Although men betray justice, God does not: murdered innocence, and the earth that receives its blood, will be avenged (see TPJS, 379; Alma 1:13; 2 Ne. 26:3; 3 Ne. 9:5-11; Rev. 6:9-11).
47 This city is not to be confused with the City of Holiness founded by the patriarch Enoch some five centuries later; see Moses 7:19; Gen. 4:16-17.
48 See John 8:44; D&C 93:25.
49 See Moses 5:29-31, 43-55; Hel. 6:21-30; Ether 8:14-26.
50 Moses 7:13;see 6:31-34; JST, Gen. 14:30-32; Ether 12:24.
51 Methuselah remained behind when his father's City of Holiness was translated from the earth; (see Moses 7:68-8:2). This curse came in the eighth century after the Fall.
52 See Gen. 18:20-33.
53 See Gen. 19:1-110.
54 Disease and pestilence are natural concomitants of famine and are, therefore, implied in this discussion.
55 See Abr. 1:29-2:1.
56 See Gen. 26:1-5.

57 The killing of the Hivites by Levi and Simeon (Gen. 34) caused Jacob to fear: "I being few in number, they shall gather themselves together against me, and slay me; and I shall be destroyed, I and my house" (Genesis 34:30). The danger of amalgamation is seen in Jacob's seeming acquiescence to the marriage of his daughter, Dinah, to Shechem and in Judah's marriage to a Canaanite woman; see Gen. 34:1-19; 38:1-3.
58 See Ruth 1:1-5.
59 See 2 Sam. 1:21.
60 See 1 Kings 17, 18; see Hel. 10:6.
61 See 2 Kings 25; Isa. 14:30; 51:19; Jer. 11-16; 18; 21; 24; 27; 29; 32; 34; 38; 42; 44; 52; Ezek. 5-7; 12; 14; 36; Matt. 24:7; Luke 21:11; Rev. 18:8; D&C 43:25; 86:7. Famine also figures prominently in the divine judgments cited in the Book of Mormon. See, for example, 2 Ne. 1:18; 6:15; 9:3; 12:4; 10:22; Hel. 10:6; 11:4-15.
62 A case in point is the American Indian whose best lands were appropriated by the "Gentiles" in many unconscionable actions.
63 See Moses 5:56; 6:15.
64 See Moses 7:33.
65 See Moses 6:57.
66 From our vantage point the mountains and deserts are quite beautiful, even inspiring, but we must remember that we have no other world with which to contrast this earth.
67 D&C 104:13-14; cf. Luke 16:10-12; Psalms 24:1-2.

Chapter Seven

The Baptism of Earth

The history of this world was foretold before it happened. Three years prior to his death, Adam met with his righteous descendants in the valley of Adam-ondi-Ahman and, as Patriarch over the human family, "predicted whatsoever thing should befall his posterity unto the latest generation" (D&C 107:56-57).[1] His prophetic recital doubtlessly included the passing of the first world order which was to end in a global deluge about seven hundred twenty-eight years later.

This was not man's first knowledge of the Flood, Enoch had learned of it more than two hundred years earlier.[2] The occasion was a face to face meeting with the Lord on mount Simeon during which Enoch was shown in vision much that would transpire down to the end of the world.[3] He saw that after his "City of Holiness" was translated from the earth, the power of Satan would become pandemic:

> And he beheld Satan; and he had a great chain in his hand, and it veiled the whole face of the earth with darkness; and he looked up and laughed, and his angels rejoiced (Moses, 7:26).[4]

Hell's happiness is mankind's misery.

Satan and the Serpent

We saw in the creation accounts that "flesh" applied to both animals and humans. Hence, the significance of this passage: "And God looked upon the earth, and behold, it was corrupt, for all flesh had corrupted its way upon the earth" (JST, Gen. 8:17; see Moses 8:30). Together with mankind, the animal kingdom had also become degenerate or debased. That Satan had a hand in this there is no doubt, for he is a defiler and a destroyer. He must have inspired deviate behavior in certain animals. Having no other "flesh" in the beginning to corrupt, he campaigned in the animal kingdom and gained a constituency there of which the serpent was the most "subtle." His war against God on this earth began in earnest when he spoke through the mouth of the serpent:

> And now the serpent was more subtle[5] than any beast of the field which I, the Lord God, had made. And Satan put it into the heart of the serpent, (for he had drawn away many after him,) and he sought also to beguile Eve, for he knew not the mind of God, wherefore he sought to destroy the world (Moses 4:4-5).

Although our first impulse may be to brand this as myth, it is not. There was a literal serpent, it could speak, and it did become an instrument of Satan's will. "The woman," said Orson Pratt, "was overcome by the devil speaking through the serpent" (JD,1:285). George Q. Cannon agreed:

> He [Satan] could not deceive Eve—or did not deceive her—except through the means of the serpent. The serpent was willing, doubtless, to let him enter, and he spoke through the serpent. It was the mouth of the serpent, but it was the voice of Satan that beguiled the woman (JD, 26:25; see 16:317).

Recall the conversation between Eve and the serpent.[6] When the serpent approached and began speaking, Eve did not demonstrate any surprise. She was in the habit of talking to the animals. The four beasts of Revelation, chapters 4-7, literally speak, sing, and praise God. Modern revelation declares them to be individual animals representative of "the glory of the classes of beings in their destined order or sphere of creation" that were "full of knowledge" (D&C 77:3-4). Commenting on these four beasts, Joseph Smith said:

> John heard the words of the beasts giving glory to God, and understood them. God who made the beasts could understand every language spoken by them. The four beasts were four of the most noble animals that had filled the measure of their creation, and had been saved from other worlds, because they were perfect: they were like angels in their sphere (TPJS, 291-92).

Even the animal kingdom must come to judgment.

The temptation of Eve is usually and incorrectly represented with a snake wrapped around a tree. The Joseph Smith Papyri (number 5) shows her confronted by a standing, two-legged, serpent-like creature. In cursing the serpent, God used the future tense, "upon thy belly shalt thou go, and dust shalt thou eat all the days of thy life" (Moses 4:20). No longer would the serpent walk; it would crawl. If it already did so, God's judgment would have been meaningless.[7] Clearly, something profound happened in the animal kingdom at the time of the Fall and subsequent thereto, for Moses tells us what science tells us: reptiles once walked the earth. Whether the serpent belonged to the dinosaur family remains to be seen, but one thing is certain, with the exception of mostly small lizards, no reptiles walk the earth today![8] How great are Satan's powers? To what extent has he been allowed to exercise physical agency down through time in his efforts to destroy the world?[9] There is no doubt of his success among men, how was it among animals in the first stages of the world?

The Wickedness of Men

By Noah's day the thoughts of men's hearts were "evil continually" (Moses 8:22). The plethora of unmitigated wickedness prompted the Lord to tell Enoch:

> Behold these thy brethren; they are the workmanship of mine own hands, and I gave unto them their knowledge, in the day I created them; and in the Garden of Eden gave I unto man his agency; And unto thy brethren have I said, and also given commandment, that they should love one another, and that they should choose me, their Father, but behold, they are without affection, and they hate their own blood; And the fire of mine indignation is kindled against them; and in my hot displeasure will I send in the floods upon them, for my fierce anger is kindled against them (Moses, 7:32-34).[10]

However, before the arm of the Lord was revealed in watery judgement, he made one final effort to save all that would be saved. Noah and his fellow prophets were to continue to cry repentance:

> My Spirit shall not always strive with man, for he shall know that all flesh shall die; yet his days shall be an hundred and twenty years; and if men do not repent, I will send in the floods upon them (Moses 8:17; see Gen. 6:3).[11]

Consequently, Noah and his righteous sons and associates, including angels from on high, spent the one hundred twenty year period immediately preceding the Flood in raising the voice of warning to a largely contemptuous, unlistening generation. The repentance required entailed much more than the shallow reformation of a minority of men and women. Mankind was enslaved to sin on so vast and pervasive a scale as to render inadequate any reformation which did not bring it to the Savior in true contrition. It is so today. As we shall see, the world faces a certain end that no righteous minority—though it number in the millions—can forestall.[12]

Moral agency was bestowed upon God's entire spirit family so that each individual might voluntarily choose him, rather than Satan, as his or her spiritual Father. This opportunity was to be extended to them in both their first and second estates. A "third part" of this "host" chose to follow the rebellious Lucifer in his civil war against Heaven; they became his spiritual sons. Enoch beheld that many of those who had remained relatively faithful to the Father in their first estate would, in mortality, reject him in favor of the father of lies. Hence the Father's lament: "they hate their own blood" (Moses 7:33). In imitating the devil's works, they became the devil's children.[13] Having renounced the Father who gave them life, they became, by adoption, the children of death.

A common misconception about Noah's generation is that almost everyone was hopelessly wicked. Such was not the case. While we are not informed as to the number who responded favorably to the prophetic message, some surely did. Recall that Enoch saw generation after generation come upon the earth after his city was translated. Satan's power grew, angels testified, and "the Holy Ghost fell on many, and they were caught up by the powers of heaven into Zion" (Moses 7:27). Consequently, all of the righteous—those who had taken upon them the name of Jesus Christ—other than Noah's family had either died or been translated before the Deluge began. The unbelieving, but basically decent men and women, of Noah's generation fell victim to the moral sickness of their fellow men. Like a deadly disease, the Flood did not discriminate between one soul and another—death came to all mankind.

Among the candidates for the terrestrial kingdom are those honorable souls of Noah's day who rejected the prophets and perished in the Flood but therafter repented of their sins in the spirit world where they accepted the fullness of the gospel.[14] As it was in the days of Noah, so shall it be when Christ returns: there will be millions of honorable men and women on the earth who, while not numbered with the righteous, will be living lives far above the character and conduct of the truly wicked. Unlike their counterparts in Noah's generation, they will survive the fiery flood of the Second Coming.

THE GLOBAL FLOOD

The Ark

Joseph Smith stated: "The construction of the first vessel was given to Noah, by revelation. The design of the ark was given by God, 'a pattern of heavenly things'" (TPJS, 251).[15] It was built during the one hundred twenty year period (2464-2344) allotted for mankind's repentance. It measured approximately 450 feet long, 75 wide, and 45 five feet high. Its draught was probably in excess of twenty feet—hence the need for the waters to cover the highest mountains more than twenty-two feet (see Gen. 6:14-16). The prophet Moroni tells us that the eight Jaredite vessels "were tight like unto the ark of Noah" (Ether 6:7).[16]

The Rains

> And, behold, I, even I, will bring in a flood of water upon the earth, to destroy all flesh, wherein is the breath of life, from under heaven; everything that liveth on the earth shall die (JST, Gen. 6:17).

The one hundred twenty years of grace passed; the warning voices of Noah and his companions fell silent. The ark of deliverance was finished and the parent animals from the greatest to the least were led aboard.[17] Noah, his wife, their three sons and their wives followed.[18] Noah turned and looked upon the first world order for the last time— "and the Lord shut him in." The closing of the ark's portal sounded the death knell of all living things that walked or crawled upon the earth or flew through its heavens.

Were there mocking onlookers? If, as we assume, there had been previous rains, they must have been relatively gentle so that there was no precedent for even localized flooding, much less the universal deluge predicted by Noah. Surely the man was mad! But derisive laughter was soon to be washed away in cries of terror as unprecedented rains consumed the ground on which the mockers stood. These first rains of record[19] occurred almost two millennia after the Fall.

On the seventeenth day of the second month—about 1656 years after the Fall—when Noah was five hundred ninety-nine years old, "were all the fountains of the great deep broken up, and the windows of heaven were opened" (Gen. 7:11). The rains came—but not as we experience them today. No tropical downpour, however intense, can compare with them. The heavens became, as it were, a vast upper ocean, its waters crashing against the face of the planet in wave after unrelenting wave, day after day and night after night.

And the waters beneath rose up. The poised subterranean aquifers, freed from their rocky chains by the mighty tremblings and tearings of the earth, burst forth from their hiding places in her bowels to join the swelling, surging, seas.[20] The waters above and the waters beneath became as one in a confluence of testimony and judgement. Terror gripped every soul as the deluge rose up and flung itself over the land. Screams drowning screams, the whole human race cried out for deliverance with the fear-born prayers of men and women driven to seek in death the God they mocked in life. All was chaos as man and beast struggled for life in the smothering arms of the swiftly rising waters. Even those men and animals who managed to flee to higher ground were quickly buried beneath a seething, twisting cauldron of mud, falling rocks, and crashing trees. The watery juggernaut swept all before it in a convulsion of furious power. Escape! To where? Death was waiting in every direction! The waters had no single point of origin, they encompassed the globe. The wrath of God was omnipresent:

> And all flesh died that moved upon the earth, both of fowl, and of cattle, and of beast, and of every creeping thing that creepeth upon the earth, and every man: And in whose nostrils was the breath of life, of all that was in the *dry land*, died.[21] And every living substance was destroyed which was upon the face of the ground, both man, and cattle, and the creeping things, and the fowl of the heaven; and they were destroyed from the earth: and Noah only remained alive, and they that were with him in the ark (Gen. 7:21-23).[22]

Death came with merciful suddenness. The waters rose rapidly so that within little more than a month, from its lowest valleys to its highest elevations, the "dry land" disappeared beneath a shoreless sea unchallenged by even the smallest fingertip of land. Such was the Deluge.[23]

Whence the Waters?

Where did the incredible volume of water needed to immerse the entire planet and its highest elevations originate? In the Creation:

> And the Gods also said: Let there be an expanse in the midst of the waters, and it shall divide the waters from the waters. And the Gods ordered the expanse, so that it divided

the waters which were under the expanse from the waters which were above the expanse; and it was so, even as they ordered. And the Gods called the expanse, Heaven (Abr. 4:6-8; see. Moses 2:6-8).

Apparently when the physical planet was organized two vast aqueous bodies were also formed. One, the planetary ocean, was in a highly condensed, liquid state. The other was diffused in a gaseous state throughout the sidereal heavens surrounding the earth.[24] This watery shroud was apparently so thinly dispersed as to make it undetectable. The sun, moon, and stars eventually appeared much as they do today, and the skies were no less clear. Yet an invisible veil of water vapor distantly enveloped the unsuspecting planet.

But why do the creation accounts tell of waters above the atmospheric heavens? A man-conceived fantasy of earth's beginnings would hardly include such a notion; how did so odd an element find its way into the accounts of both Abraham and Moses? It was an important fact the Lord wanted included in the divine narratives—one that was not only essential to the basic details of the earth's birth, but of its rebirth as well. These waters were prepared in anticipation of the Flood which was programmed from the beginning. Every critical event that was to take place on this planet throughout its temporal career was incorporated into its design and organization by its Architects.

Reality of Flood

> And the flood was forty days upon the earth; and the waters increased and bare up the ark, and it was lifted up above the earth. . . .And the waters prevailed exceedingly upon the earth; and all the high hills, that were under the whole heaven were covered. Fifteen cubits upward did the waters prevail; and the mountains were covered (Gen. 7:17, 19-20).

The psalmist wrote: "Thou coveredst it with the deep as with a garment: the waters stood above the mountains" (Ps. 104:6).

While most geologists reject Moses' account of a global deluge outright, there are others who simply edit it down to a regional affair.[25] However, Moses' account, combined with other inspired testimony, simply do not allow for such a "wresting" of Scripture.[26] Contrary to the thinking of most higher critics of the Bible, the ancient prophets were not naive simpletons. A local flood would hardly have covered the highest mountains from twenty-two to twenty-seven feet! Nor would it have required five months to the day for the waters to recede sufficiently for the ark to ground itself on a mountain. Thereafter, on the first day of the tenth month, the tops of the mountains finally reappeared. (see Gen. 7:17-24).[27]

Regardless of the accuracy of the details found in Genesis, no one can honestly question Moses' intent to describe the Flood as an unprecedented world-wide phenomenon. The Book of Mormon testifies to the correctness of Moses' account by declaring "after the waters had receded from off the face of this land it [the Western Hemisphere] became a choice land above all other lands" (Ether 13:2).

It would have been ludicrous for Noah to build, or Moses to describe, so large a vessel in terms of a regional disaster. The ark's size was dictated primarily by its animal cargo. Why preserve a few representative animals of one limited locality when the same species were likely to be found virtually everywhere else? And why would Noah—who had been warning of the impending catastrophe for one hundred twenty years—build even a rowboat rather than remove his family to higher ground or to another land if he knew that a given area was going to be flooded?

Noah was not insane, he was a prophet of God, who "stands next in authority to Adam in the Priesthood" (TPJS, 157; see

Luke 1:19, 26). He is Gabriel, the angel who appeared to Zacharias in the temple and to Mary in Nazareth. Would such a man have been so destitute of reason as to be incapable of distinguishing between a global deluge and a regional flooding? Would Moses? Writing of those in the last days who would mock the belief in Christ's return, the apostle Peter compared that future world-wide event with an earlier world-wide event, the Flood:

> For this they willingly are ignorant of, that by the word of God the heavens were of old, and the earth standing out of the water and in the water: Whereby the world that then was, being overflowed with water, perished: But the heavens and the earth, which are now, by the same word are kept in store, reserved unto fire against the day of judgement and perdition of ungodly men (2 Pet. 3:5-7).

Had the Flood not been universal, the validity of Peter's argument would be totally lost. But Peter's point is clear: just as the first world order ended in water, so will the second world order end in fire. As we shall see, water and fire are critical elements in the salvation of Earth and mankind.

And most importantly, if the Flood was not world-wide, God has violated his oath, his immutable word—given first to Enoch and reconfirmed to Noah—that this planet would "never more be covered by the floods" (see Moses 7:50-51; Gen. 9:15). For if the Flood was not global, than it was limited, and if it was limited, then the Lord has foresworn himself since limited floodings have occurred throughout recorded history. However, the Lord did not violate his oath. His promise to never again "destroy all flesh" by a global deluge did not preclude the loss of some life in subsequent floods of a localized nature. God's oath-bound assurances to Enoch and Noah, combined with the fact that area floodings continue to occur, rule out the possibility that the Noachian deluge was but another in a series of ongoing natural disasters. However, he who organized this planet and covered it with an unheard of global sea from whence its land masses eventually emerged, was fully capable of returning it to that watery state at his pleasure. For mankind, the Flood was a miracle.

Sequence of the Flood

(This chart summarizes information in Genesis 7 and 8.)

Event	Date	Elapsed Days
1. Noah enters Ark Rains begin (7:11, 13)	600th Year of Noah 2nd mo., 17th d.	0
2. Heavy rains end. Ultimate water level reached (7:4, 12, 17-20, 24)	3rd mo., 27th d.	40*
3. Wind rises. Waters begin to recede Ark grounds on mountains of Ararat (7:24; 8:1-4)	7th mo., 17th d.	110
4. Mountain tops begin to reappear (8:5)	10th mo., 1st d.	74
5. Raven released, does not reenter Ark (8:6-7)	11th mo., 11th d.	40*
6. Dove released, returns (8:8-9)	11th mo., 18th d.	7
7. Dove released, returns w/olive leaf (8:10-11)	11th mo., 25th d.	7
8. Dove released third time	12th mo., 2nd d.	7
9. Waters completely recede Ark's covering removed (8:13)	1st mo., 1st d.	29
10. Earth dry; Noah leaves Ark (8:14)	2nd mo., 27th d.	56
	Total time in Ark:	370

A thirty day month is assumed. If this was a solar year, the total would be 375 days.
*While the number "40" is sometimes interpreted symbolically, in this instance it appears to be used literally.

In summation, the rains fell and the waters rose for forty days, maintaining their level for another one hundred ten days—one hundred fifty days in all. The global sea then began to recede, causing the Ark to ground itself upon the mountains of Ararat in Armenia. Within seventy-four days thereafter the mountain tops began to reappear. Ninety days later the waters had completely receded, leaving a soggy earth. After another fifty-six days, Noah and his family walked on dry ground for the first time in more than a year.[28]

Ancient Mountains

It is admittedly difficult for us to envision water rising high enough to cover a mountain such as Everest, the summit of which is nearly six miles above the surface of the sea! However, the submersion of the highest mountains might not have required as great a miracle or volume of water in Noah's time as would be required now.[29] The extent to which the "dry land" had emerged from the earth's primordial ocean prior to the Flood is unknown. Also unknown is the maximum elevation above sea level of any portion of that land. The assumption that the continents and their mountain ranges were much the same millions of years ago as they are today is only that, an assumption.[30]

There is strong scriptural evidence that the planet's major mountain ranges (such as the Himalayas, the Andes, the Alps, and the Rockies) did not exist when it was first organized, nor, in all likelihood, before the Flood. The first was as the last shall be. Prophecies concerning conditions in the end of time are actually revelations of conditions in the beginning of time. They indicate that while there may have been some relatively low mountains, the planet was characterized by rolling hills and broad valleys.

In connection with "the restitution of all things," the tenth Article of Faith states that "the earth will be renewed."[31] This renewal or restoration includes the lowering of the mountains with the consequent raising of the valleys. Thus, the planet's present high mountain ranges will no longer exist, being no longer necessary from either a spiritual or a physical standpoint. In a sense, they are blemishes on the earth's face symbolizing its fallen, divided, condition; when it rises to a terrestrial state, these "blemishes" will no longer be appropriate.

The term, "mountain," is quite relative and has been applied to large hills as well as towering peaks.[32] For example, both the traditional "mount of temptation" (west of Jericho) and the traditional site of the Sermon on the Mount (northwest of the sea of Galilee) are scarcely above sea level. Indeed, there are very few scriptural references to high mountains. The book of Ether tells of the brother of Jared quarrying sixteen stones from a mountain "called the mount Shelem, because of its exceeding height" (Ether 3:1). Nephi writes of being caught away in the Spirit of the Lord "into an exceedingly high mountain, which I never had before seen, and upon which I never had before set foot" (1Ne.11:1). Neither of these mountains is identified, so there is no way to determine just what was meant by "exceedingly high." Moses was "caught up into an exceedingly high mountain" when the Lord God commissioned him to return to Egypt and deliver Israel from slavery.[33, 34] The parallel account found in Exodus identifies the mountain as Horeb (or Sinai)—"the mountain of God" (Ex. 3:1).[35] Horeb is thought to be located somewhere in the Sinai peninsula (the precise location is conjectural), where the mountains—which vary in height from about 6700 to 9000 feet above sea level—rise only 1800 to 4000 feet above the desert floor. While such mountains may have been considered "exceedingly high" by men whose lives were largely spent in the desert's hills and valleys, they would hardly compare with ranges in other parts of the world having elevations approaching thirty thousand feet.

Most importantly, in every instance where a mountain is described as being "exceedingly high," the passage pertains to events *after the Flood and the division of the earth*.[36] In all likelihood, the great ranges of the world were thrown up in astonishingly brief mountain-building periods partly concurrent with and partly subsequent to the Flood. Such profound topographical changes would have been entirely consistent with and reflective of God's division of the planet whereby he separated the human family spiritually, linguistically, and geographically.

The time required to bring about major geologic changes is dependent upon the forces involved. When the Almighty exercises his power, time ceases to be a factor. For example, in a matter of hours, a "great mountain" was formed to occupy the site where the Nephite city of Moronihah had stood.[37] The death of Christ was accompanied by other extensive changes in the character and topography of portions of the Western Hemisphere.

Then too, even discounting the foregoing and assuming that the mountains of Noah's time were of formidable height, it would have been a simple matter for the Creator to cause Earth's land masses—which are said to float upon cushions of plastic subterranean material ("plates")—to partially sink beneath the flood waters and then rise again even as the dry land emerged from the global sea on the second day of creation.[38] Thus, Earth—as with any recipient of baptism—would have been an active participant in its own immersion. Baptism is not employed as simply a convenient metaphor to argue the case for a universal deluge: Its use is deliberate. The Flood is a witness for baptism and baptism is a witness for the Flood.[39]

The Flood Was Justified

While men are free to destroy themselves, they are not free to destroy others (see Alma 36:9). The contagious nature of the evil practices of Noah's generation took an unacceptable toll among the innocent and the young. To allow a world ripened in iniquity to continue to exist would have been a mockery of justice and an inevitable defeat for the unborn. John Taylor defended God's moral integrity in sweeping the race from the earth, explaining that it was done for the eternal welfare of both the living and the unborn:

> ...the Lord as a great cosmogonist, took in the various stages of man's existence, and operated for the general benefit of the whole. But was it not cruel to destroy them? I think God understood precisely what He was doing. They were His offspring, and He knowing things better than they did, and they having placed themselves under the power and dominion of Satan, He thought they had better be removed and another class of men be introduced. Why? There were other persons concerned besides them. There were millions of spirits in the eternal worlds who would shrink from being contaminated by the wicked and corrupt, the debauchee, the dishonest, the fraudulent, the hypocrite, and men who trampled upon the ordinances of God. It might seem harsh for these men to be swept off from the face of the earth, and not allowed to perpetuate their species thereon; but what about the justice of forcing these pure spirits to come and inhabit tabernacles begotten by debauched corrupt reprobates, the imagination of whose heart was only evil, and that continually—what about them? Had they no rights that God was bound to respect? Certainly they had, and He respected them. He cut off the wicked (JD, 26:35).[40]

Adam's world had to die so that Noah's world might live.[41] However, the Almighty was also mindful of the salvation of the wicked; divine mercy seasoned divine justice. As Orson F. Whitney observed:

God does not punish except to save, He never chastens except to purify. In sweeping the antediluvian races from the earth, it was an act of mercy to them, that they might not add sin to sin and heap up iniquity until they could not have been pardoned. He swept them off when their cup was full, and imprisoned their spirits while their bodies mouldered in the grave. Jesus, however, while His body was lying in the tomb, went and preached to the spirits in prison; those who rejected the message that was offered to them by Noah, and were swept away by the flood (JD, 26:268; see D&C 138:29).

We must not forget, the judgments of God are largely self-imposed; we dictate our own eternal fate. George Q. Cannon warned: "If you are damned, you damn yourselves; you will be the instrument of your own damnation" (JD, 26:250; see TPJS, 357). The Lord's responses to human behavior are but the harvest of man's willl. He, as all parents should be, was free of the blood and sins of his children who perished in the Flood.

Earth's Baptismal Covenant

And again, verily I say unto you, the earth abideth the law of a celestial kingdom, for it filleth the measure of its creation, and transgresseth not the law (D&C 88:25).

The law it obeys is the law of Christ, the new and everlasting covenant by which men and worlds are sanctified in immortal glory (see D&C 88:21). Basic to that covenant law is the ordinance of baptism—the two-part ordinance corresponding to the dual nature of a living soul—the spirit joined to its physical counterpart.[42] Baptism both symbolizes and serves the dual nature of a living soul in that it involves the immersion of the physical body in water and the subsequent immersion of the spirit body in the Holy Spirit. For this reason Jesus told Nicodemus: "Except a man be born of water *and* of the Spirit, he cannot enter into the kingdom of God" (John 3:5). The law applies with equal force to Earth: it cannot enter the celestial order of kingdoms without being born again of water and of the Spirit.

Like Jesus, Mother Earth had to be baptized to fulfill all righteousness. And, like Jesus, she was baptized for her perfecting, not for her sins. Such was not the case with her human children; mankind had defiled her with their abominations. In the wisdom of God it was decreed that she should be freed of her human burden so that both she and her unborn children might have a new beginning. Orson Pratt said Earth was baptized by Christ himself:

> The first ordinance instituted for the cleansing of the earth, was that of immersion in water; it was buried in the liquid element, and all things sinful upon the face of it were washed away. . . .As man cannot be born again of water, without an administrator, so the earth required an agency independent of itself, to administer this grand cleansing ordinance, and restore it to its infant purity. That administrator was the Redeemer himself (JD, 1:331).[43]

Brigham Young bluntly stated, "the filthiness that has gone forth" out of the earth is "You and I, and all the inhabitants of the earth. . . .it was baptized for the remission of sins." He continues:

> The Lord said, I will deluge (or immerse) the earth in water for the remission of the sins of the people; or if you will allow me to express myself in a familiar style, to kill all the vermin that were nitting, and breeding, and polluting its body; it was cleansed of its filthiness; and soaked in the water, as long as some of our people ought to soak. The Lord baptized the earth for the remission of sins, and it has been once cleansed from the filthiness that has gone out of it, which was in the inhabitants who dwelt upon its face (JD, 1:274).

The very nature, purpose, and symbolism of baptism necessitates the total immersion of the entire body; sprinkling or pouring water on the head or some other part of the body is not baptism (see DS, 2: 320). The ordinance signifies the remission of all sins via a total immersion in the blood of Christ. Other modes of baptism violate its divine symbolism.[44] This fact alone should be enough to convince any one having an enlightened understanding of the Scriptures—ancient and modern—that the waters of the Flood covered every particle of land on the globe. This is accepted LDS doctrine.[45] At its physical birth, the "dry land" of Mother Earth came forth from the womb of a global sea only to return to that womb via the Flood. She was baptized in her childhood and born again of water. She had received the first ordinance of salvation in the kingdom of God at the hands of her Creator-Savior.

The Sign of an Oath

Having once received the ordinance of baptism in water, the earth was never again to require that ordinance. The Flood, like the Atonement, would not be repeated. Enoch, beholding in vision the future deluge, wept bitterly and pled with the Lord "that the earth might never more be covered by the floods." The Lord swore to Enoch with an oath that such would be the case (see Moses 7:50-51). Isaiah wrote that the Redeemer's forgiveness of Israel was as certain as his oath-bound promise to never again flood all the earth:

> For this is as the waters of Noah unto me: for as I have sworn that the waters of Noah should no more go over the earth; so have I sworn that I would not be wroth with thee, nor rebuke thee (Isa. 54:9; see 3 Ne. 22:9).

Following the Flood, the Lord comforted Noah—and his posterity down through time—with the same assurance: "And I will establish my covenant with you; neither shall all flesh be cut off anymore by the waters of a flood; neither shall there any more be a flood to destroy the earth" (Gen. 9:11; see 8:21; 9:15). The sign of that immutable promise was to be the rainbow—a token which was to join all of the other signs the Lord had previously appointed for man's comfort and guidance.

> And God made a covenant with Noah, and said, This shall be the token of the covenant I make between me and you, and for every living creature with you, for perpetual generations; I will set my bow in the cloud; and it shall be for a token of a covenant between me and the earth (JST, Gen. 9:18-19).

Through Joseph Smith, God not only confirmed the reality of the Flood, he also confirmed Moses' testimony that the rainbow originated "in the days of Noah" (TPJS, 340). We assume that the presence of water vapor and sunlight always has the potential to create a rainbow, yet we are told that no rainbow, as such, had ever arched the heavens. Had the rainbow been a common phenomenon, it would not have been employed as a unique sign. The rainbow—which signifies security from a global deluge—would hardly have been observed prior to such a deluge! The night of God's justice was signified by the Flood, the day of his mercy by the rainbow. As that day passes and the sun sets, mercy will cease and the rainbow will not be seen; darkness will creep across the world and the blessed rains—which the Lord has patiently sent upon the just and the unjust—will cease.

The Aftermath

Almost four months after the flood waters began to recede, Noah began a series of probings to ascertain conditions beyond the immediate vicinity of the ark. His scout was the raven, a strong, indelicate bird eminently suited for the assignment. As Noah doubtlessly expected, the raven never returned. It found refuge on the emerging mountains from whence it could fly forth in search of carrion and the like. It was a bird of death that lived off of the dead. A week later Noah released a dove, only to have it quickly return to the safety of the ark. Upon being released the following week, it returned again; but this time with an olive leaf in its beak. Noah knew that all would be well. The olive tree, so hardy and enduring, so capable of surviving under severe conditions of abuse and neglect, was coming alive.

How appropriate were the signs! The olive tree was designated the divine symbol of Israel, the Father's family in time and eternity. From it sprang the olive leaf—the symbol of the Lord's message of peace and salvation which Israel was chosen and ordained to bring to the nations of the world.[46] And just as the dove brought the olive leaf to Noah, so was the message of salvation through Jesus Christ to be brought to mankind by the power of the Holy Ghost—of which the dove is the divine and incorruptible symbol.[47] How marvelous are the ways of the Lord! What has been long regarded as the incidental trappings of an ancient tale prove to be nothing less than symbolic prophecies of God's post-Flood labors in bringing to pass the salvation of mankind!

Upon being released for the third time the dove did not return. There was no need to do so. Hope was no longer confined to the ark. The dove's flight back into the world signaled the return of the Spirit of God to Earth. The Lord had not forsaken his children; life and peace were again proffered them. And so the second world order began with eight human souls.[48] Noah and his family left the ark and entered a rain-washed world where all was fresh and clean and promising. The animals were released to run upon the land, the birds to restore life and song to the heavens. Ancient words were repeated as God renewed the commandment: "Be fruitful, and multiply, and replenish the earth" (Gen. 8:17; 9:1, 7). Life was to be given a second chance; man and creature were to go on.

But a shadow fell across the day; a portent marred the moment. Whereas man, through Adam, had been designated the steward and protector of lesser life in that lost Eden, now man, through Noah, became its enemy and destroyer:

> And the fear of you and the dread of you shall be upon every beast of the earth, and upon every fowl of the air, upon all that moveth upon the earth, and upon all the fishes of the sea; into your hand are they delivered. Every moving thing that liveth shall be meat for you; even as the green herb have I given you all things (Gen. 9:2-3).[49]

If Adam's world—founded upon the highest principles of truth, mercy and peace—eventually fell to ignominy before a satanic assault, what real hope was there for Noah's world when its very birth was a product of death and its infant beginnings were stained by blood? Truly, there had been a descending; the higher law of reverence for life had been replaced by a lesser law of conflict and death.

Would man limit the taking of life to the animal kingdom? Would he stay his hand from striking down his fellows? He had not done so in a better day, would he do so now? The distinction between the greater and the lesser life became so blurred in the minds of men as to somehow merge all life into one life. Knowing it, the Lord testified to the sanctity of all life by making men accountable for all life:

> And surely, blood shall not be shed, only for meat [food], to save your lives, and the blood of every beast will I require. And whoso sheddeth man's blood, by man shall his blood be shed; for man shall not shed the blood of man. For a commandment I give, that every man's brother shall preserve the

life of man, for in mine own image have I made man (JST, Gen. 9:11-13; see D&C 49:21).

Another fall had taken place; the planet was not what it had been in the beginning, nor even what it was when Adam and Eve left the Garden. It was a new, but not a better, world. For one thing the longevity of man was greatly diminished. Whereas the antediluvian patriarchs lived for many centuries, the lifespan of those enumerated in Scripture rapidly declined thereafter. For example, the average lifespan of the nine patriarchs (Enoch being excluded) from Adam to Noah was 912 years, while that of the first nine patriarchs born after the Flood was 285 years—a sixty-nine percent drop in life expectancy.

Why so drastic a decline? No one knows. Even the eating of meat has been suggested as a possible contributing cause. However, a more likely explanation may be found in the drastic modification of the environment which attended the Flood. Prior to it, the upper waters may well have served as a protective shield against the cosmic and solar rays which constantly bombard the earth's surface. The removal of that screen—in whole or in part—exposed the earth and its denizens to higher levels of radiation thereafter with a resulting decline in longevity more or less proportionate to the increased amounts of natural radiation reaching the earth.[50] Thus, even as the Fall set in motion those processes of change and decay which inevitably end in death, the Flood—with its attendant modification of the global environment—greatly, and necessarily, accelerated those self-same processes.

The "heavens and the earth, which are now" await the coming of creation's Redeemer. He will complete the saving ordinance begun in the days of Noah. So that while the "wisdom of the world" relegates the immersion of the earth in baptismal waters to the realm of fantasy, it was nonetheless a magnificent reality, an essential act of preparation for even more wondrous things to come.

NOTES

1. This assembly took place about 3073. Joseph Smith beheld it in vision: "I saw Adam in the valley of Adam-ondi-Ahman. He called together his children and blessed them with a patriarchal blessing. The Lord appeared in their midst, and he (Adam) blessed them all, and foretold what should befall them to the latest generation" (TPJS, 158). A similar conference will be held at Adam-ondi-Ahman preparatory to the world coming of Jesus Christ. See D&C 116; Dan. 7:9-10 and TPJS, 157.
2. Joseph Smith wrote that God's seers and prophets "saw the flood before it came" (TPJS, 12).
3. See Moses 7:2-3. Enoch was born 622 years after the Fall. He was 365 years old at the time of his meeting with Adam; his "City of Holiness" had been established for 240 years.
4. The "chain" represented the "chains of hell," the sins and spiritual blindness which bind men's souls to the devil; see Alma 12:10-11.
5. "More subtle," not most subtle. This suggests a difference between the serpent and any creatures God had made. Did Satan create (or re-create) the serpent? He has been credited with such powers as when the Egyptian magicians turned their rods into serpents; see Ex. 7:10-12; JD, 2:12.
6. See Gen. 3:1-6.
7. Was this serpent a small dinosaur? Was the fate of all dinosaurs hidden in God's judgment of the serpent? And was that judgment carried out in the Flood?
8. Almost all lizards have four feet with five toes on each foot. There are over 3000 species throughout the world.
9. See D&C 121:4.
10. Enoch may have been the first mortal to learn of a future world-wide flood since this is the most ancient extant scriptural reference to it; see Alma 10:22. Note the play on words: only the flood waters could cool God's "hot displeasure."
11. This statement has been erroneously interpreted as referring to man's life span, that it would be one hundred twenty years. It simply refers to the length of time granted to Noah's generation to repent. It is called a "prophetic generation."
12. See Matt. 24:34-41; D&C 5:18-19; 33:2-4; 38:11-12; 86:4-7.
13. See John 8:39-44.
14. See Moses 7:38-39, 57; 1 Pet. 3:18; 4:6; D&C 76:73-75; 88:89; 138:7-10, 28-35.
15. Orson Pratt opined: "He [Noah] had a Urim and Thummim by which

he was enabled to discern all things pertaining to the ark and its pattern" (JD, 16:50).
16 See Gen. 8:4. Some claim its remains have been located on mount Ararat in Turkey, but this is unverified.
17 The animals Noah saved were of both the clean and unclean varieties; see Gen. 7:1-3, 8- 9, 14-15.
18 The lineage of Cain was represented on the ark by Egyptus, the wife of Ham. The record indicates that Ham was a righteous man, a son of God. Egyptus was likewise a worthy woman whose grandson, Pharaoh, was also a righteous man;see Moses 8:13; Abr. 1:21-26.
19 There is no scriptural reference to rain up to that time other than the statement that in the morn of creation "there went up a mist from the earth, and watered the whole face of the ground" (Gen. 2:6; see Moses 3:6; Abr. 5:5-6).
20 It has been suggested that the withdrawal of the Spirit of the Lord triggered the Flood: "When that Spirit was partly withdrawn in the days of Noah, the immense masses of water confined below in 'the fountains of the great deep' broke their fetters and overflowed the Earth, while, at the same time, the vapors in the atmosphere condensed and fell down in torrents as if the 'windows of heaven' had been opened (Gen. 7:11), and the result was the destruction of every living creature, except those in the ark" (Smith and Sjodahl, 380).
21 Marine life, as such, was not destroyed, but much will be hereafter.
22 Although Noah filled the ark with a host of creatures representative of the animal kingdom, not every prediluvian species was destined for a place in the second world order. Some creatures—such as those giant reptiles comprising the dinosaur family of the so-called Mesozoic period, either failed to survive the Fall or were swept from the earth in the cataclysmic events associated with the Deluge.The many species known today comprise the posterity of those animals preserved by Noah. The assumption that he gathered literally every creature from the merest insect to the largest mammoths is unwarranted. Within the micro-evolutionary bounds set by the Almighty—who is also capable of "genetic engineering"—the surviving classes have doubtlessly proliferated into a number of species and subspecies via genetic modifications and cross-breeding; see JST, Gen. 9:2.
23 See JST Gen. 7:19-20; 8:5.
24 Hydrogen, the primary element in water, is the most common of all known elements.
25 The Flood is treated as a myth by most commentators who trace its origin to much earlier "Mesopotamian originals." The Genesis version is viewed as a combination of materials from two different Hebrew sources: the "P" (Priestly) and the "J" (Yahwist). While it is true that

the Babylonian Gilgamesh Epic and, to a lesser extent, the Sumerian legend contain a number of basic elements surprisingly similar to those found in Genesis, overall they are clearly distorted heathen adaptations of the original scriptural account which may have antedated Moses by a thousand years. Their value does not lie in their content but in their very existence. The fact that the basic concept of a global flood was widespread in the post-diluvian world is a strong argument for its authenticity.

26 Key scriptural references to the Flood are Isa. 54:9; Matt. 24:37-38; Alma 10:22; 3 Ne. 22:9; Ether 6:7; 13:2.

27 "Many ancient sources recall that after the waters of the Flood had subsided there came a great 'Windflood' which converted large areas of the world to sandy deserts; A. Haldar considers the Sumerian version of the Windflood to be 'an excellent example of a text describing historical events in terms of religious language.' The historical reality is attested by windblown sand deposits from various and widely separated periods, which can be broadly correlated with some of the major migrations of peoples" (Nibley, 10: 33-34).

28 Noah observed his 600th birthday during the Flood. He lived another 350 years.

29 John A. Widtsoe wrote: "All parts of the earth were under water at the same time. In some places the layer of water might have been twenty-six feet deep or more; in others, as on sloping hillsides, it might have been only a fraction of an inch in depth" (*Evidences and Reconciliations*, 111). In other words, technically speaking, a torrential rain would have been sufficient to cover everything with a thin layer of water at any given moment. But would it have destroyed all life? Gen. 8:5 states that "the tops of the mountains" were not seen until over eight months after the rains began or six months after they ended.

30 Fossil evidence indicates that "all the major mountain ranges of the present world evidently were uplifted within the most recent eras of geologic history" (Morris and Whitcomb, *The Genesis Flood*, 142). Geologist now believe that the continents ride on about twenty-four "tectonic plates." It is believed these "plates" account for "continental drift"—thought to be a major cause of the mountain building process. The period of the Flood, including the division of the earth into separate hemispheres and continents, was probably the earth's great mountain-building period. If the continents began shifting in the days of Peleg, the great mountain ranges can only be a few thousand years old.

31 Speaking of this, Parly P. Pratt wrote: "When a Prophet speaks of the restoration of all things, he means that all things have undergone a change, and are to be again restored to their primitive order, even as they first existed" (*Voice of Warning*, 85).

32 It has been applied to manmade structures. For example, the ziggurats (of which the tower of Babel is the best known) of ancient Mesopotamia were artificial mounds of earth thrown up to simulate mountains and were, in fact, designated as such: "The significance of the ziggurats is revealed by the names which many of them bear, names which identify them as mountains. That of the god Enlil at Nippur, for example, was called 'House of the Mountain, Mountain Storm, Bond between Heaven and Earth'" (Frankfort, 54).
33 See Moses 1:1, 2, 26)
34 The phrase "an exceedingly high mountain" may also be a metaphor for the presence of the Lord. Moses 1:46 tells us that the name of the mount on which Moses stood "shall not be known among the children of men." God is a mountain Man and where he is, there is a spiritual mountain. His temple is a symbolic mountain: "the mountain of the Lord's house shall be established in the top of the mountains" (Isa. 2:2). Zion becomes "mount Zion" when it is glorified by the presence of the Lord; see Isa. 8:18; 55:20; D&C 76:66; TPJS, 340.
35 It has been suggested that the names Sinai and Horeb are not to be identified with any one mountain, being conceptual terms denoting emptiness or desolation.
36 Enoch, who lived before the flood, talked with God on mount Simeon, but no reference is made to its height. See Moses 7:2-4.
37 See 3 Ne. 8:10.
38 As before stated, current geophysical research indicates that the relatively thin outer crust of the earth (comprising the continents and ocean basins) is supported by a number of massive, moving plates of layered materials which are, in turn, floating on the molten matter, called magma, found in the core or mantle of the earth.
39 The apostle Peter understood this point and equated the ark which saved Noah's family from physical death to baptism which saves us from spiritual death. See 1 Pet. 3:20-21.
40 See JD, 24:290-91; 17:205-06; 18:330-31; 19:158-59.
41 See 2 Pet. 3:5-7.
42 See D&C 88:15.
43 See also JD, 1:291-93; 16:313-14; 21:323-24. While Orson Pratt calls baptism "the first ordinance instituted for the cleansing of the earth," its validity depended upon the atonement. Since the atonement retroactively freed both Earth and mankind of the burden of original guilt, it may be said that it was the the Savior's own ordinance of salvation. Thus, as Orson Pratt observed, "Both man and the earth are redeemed from the original sin without ordinances" (JD, 1:291).
44 See John 3:23; Rom. 6:4; D&C 20:72-74; Morm. 9:6; Moses 6:64; TPJS, 12, 198, 360.

45 See *Evidences and Reconciliations,* 127-28; *Mormon Doctrine,* 289; JD, 1:274, 331.
46 Referring to D&C 88, the Prophet Joseph Smith wrote W.W. Phelps: "I send you the 'olive leaf' which we have plucked from the Tree of Paradise, the Lord's message of peace to us" (TPJS, 18).
47 Joseph Smith said: "The sign of the dove was instituted before the creation of the world, a witness for the Holy Ghost, and the devil cannot come in the sign of a dove" (TPJS, 276).
48 See 1 Pet. 3:20.
49 Man and beast were herbiverous before the fall (see Gen. 1:29-30). While divine authorization to eat flesh may have been given prior to Noah's time, this passage contains the first scriptural approval for so doing.
50 Research into the effects of radiation at all levels of intensity, no matter how sublethal, has demonstrated that radiation does "appreciably reduce the life span." Commenting on this fact, Morris and Whitcomb wrote: "If such effects can be observed in a short lifetime as a result of artificial radiations, it is certainly possible that much greater effects on longevity would have been produced over the millenniums by the natural background radiation" (*The Genesis Flood,* 401).

Chapter Eight

Earth is Divided

Mother Earth has grown old, although still beautiful to mortal eyes, the magnificence with which she was originally endowed by her Creator is but dimly perceptible now. She has been shorn of her pristine glory, not only because of the Fall and the Flood, but also because she has experienced dismemberment and division. As we shall see, certain regions of the earth have been radically modified while others have been literally severed from the planet and relocated elsewhere in the sidereal heavens.

The Primal Land

On the third day of creation, a solitary land mass rose up out of Earth's primordial waters.[1] The extent of this "dry land" is not given, but it probably approximated that of the present continents.[2] However, subsequently additional land emerged from the global sea. Less than a thousand years after the Fall in the days of Enoch:

> There also came up a land out of the depth of the sea, and so great was the fear of the enemies of the people of God, that they fled and stood afar off and went upon the land which came up out of the depth of the sea (Moses 7:14).

Enoch's City of Holiness

Enoch, the seventh patriarch,[3] was blessed with a wise and righteous father who taught him "in all the ways of God" (Moses 6:21). At the age of twenty-five, this remarkable young man was ordained a patriarch (an evangelist) by Adam himself.[4]

At about that same time, he responded to a divine call to take the message of repentance to the outside world with such zeal that he eventually established a church of God.[5] This great prophet, seer and revelator was mighty in word and deed:

> And so great was the faith of Enoch, that he led the people of God, and their enemies came to battle against them; and he spake the word of the Lord, and the earth trembled, and the mountains fled, even according to his command; and the rivers of water were turned out of their course; and the roar of the lions was heard out of the wilderness; and all nations feared greatly, so powerful was the word of Enoch, and so great was the power of the language which God had given him.(Moses 7:13; see JST, Gen. 14:30, 31).

The "church of Enoch" grew in both numbers and righteousness to such an extent that, at age sixty-five, forty years after beginning his ministry, he formally established "the City of Holiness, even ZION."[6] It was a spiritual ark whereon the antediluvian saints found safety from the wickedness and warfare of the outside world. God "dwelt with his people" thus transforming the city into a microcosm of Christ's world-wide millennial reign (see Moses 7:16). So glorious was the City of Holiness, Enoch thought it would "dwell in safety forever" (Moses 7:20). Zion was to endure, but not on this earth; it was to dwell with Christ on another sphere.[7] The righteousness of Zion could not save a world drowning in its own carnality. Indeed, the moral degeneracy of the children of men made it impossible for the city to long remain among them. Wilford

Woodruff said that Enoch's Zion did not remain on the earth because "wickedness prevailed." He explained:

> . . .the majority of the human family in that generation were wicked; they were not ruled over by the Lord; and, hence, there were not men enough on the face of the earth, in that generation, who were willing to receive the Gospel, keep the commandments of God, and work the works of righteousness, for Enoch to have power to remain on the earth (JD 11:242; see 26:34).

Joseph Smith wrote:

> He [God] selected Enoch, whom He directed, and gave His law unto, and to the people who were with him; and when the world in general would not obey the commands of God, after walking with God, he translated Enoch and his church, and the Priesthood or government of heaven was taken away (TPJS, 251).

And so was a portion of Earth!

> And Enoch and all his people walked with God, and he [God] dwelt in the midst of Zion; and it came to pass that Zion was not, for God received it up into his bosom; and from thence went forth the saying, ZION IS FLED (Moses 7:69).

The reality of this unprecedented miracle is confirmed in the New Testament:

> . . .by faith Enoch was translated that he should not see death; and was not found, because God had translated him: for before his translation he had this testimony, that he pleased God (Heb. 11:5; see Gen. 5:22).

After three hundred sixty-five years, the City of Holiness, was caught up from the earth, lifted into the heavens, and translated to a *terrestrial* sphere where Enoch and his people continue to walk with Jesus Christ who declared: "I am the same which have taken the Zion of Enoch into mine own bosom" (D&C 38:4). That the Lord's "bosom" is not located on this globe was made explicit by Christ when he identified Enoch and his brethren as those "who were separated from the earth, and were received unto myself" (D&C 45:12).[8] Remaining in a state of translation, they are with him still.[9]

Righteous individuals were translated to Enoch's city *after* it was removed from Earth: "The Holy Ghost fell on many, and they were caught up by the powers of heaven into Zion" (Moses 7:27). These worthy souls were also numbered with the "general assembly and church of Enoch" gleaned from the nations of the first world order.[10] In her previously quoted poem, "Address to Earth," Eliza R. Snow writes of this unprecedented event:

> When Enoch could no longer stay
>   Amid corruption here,
> Part of thyself was borne away
>   To form another sphere.
>
> That portion where his City stood
>   He gain'd by right approv'd;
> And nearer to the throne of God
>   His planet upward mov'd.

Wandle Mace, a faithful friend of Joseph Smith, quoted him as saying that "when Enoch and his City was taken away, a portion of the Earth was taken, and it would again be restored."[11] Not only were Enoch and his people translated from the earth, but their city literally fled into the heavens![12] Zion was not limited to a few square miles of urban streets and buildings; it encompassed a vast rural hinterland as well; for "the Lord blessed the land, and they were blessed upon the mountains, and upon the high places, and did flourish" (Moses 7:17). According to Brigham Young, in departing, they left nothing behind:

> . . .he obtained power to translate himself and his people, *with the region they inhabited,* their houses, gardens, fields, cattle, and all their possessions. He had learned enough from Adam and his associates to know how to handle the elements, and those who would not listen to his teachings were so wicked that they were fit to be destroyed, and he obtained power to take his portion of the earth and move out a little while, *where he remains to this day* (JD, 3:320; emphasis added; see 8:279).

We may judge something of the dimensions of Enoch's city from its location which was apparently revealed to the Prophet Joseph Smith. Joseph Young (brother of Brigham Young) summarized what he had heard Joseph Smith teach in a discourse given in Nauvoo:

> That the people, and the city, and the foundations of the earth on which it stood, had partaken so much of the immortal elements, bestowed upon them by God through the teachings of Enoch, that it became philosophically impossible for them to remain any longer upon the earth; consequently, Enoch and his people, with the city which they occupied, and the foundations on which it stood, with a large piece of earth immediately connected with the foundations and the city, had assumed an aerial position within the limits of our solar system; and this in consequence of their faith.[13]

The Prophet, said Joseph Young, further stated, "the City of Enoch would again take its place in the identical spot from which it had been detached, now forming that chasm of the earth, filled with water, called the Gulf of Mexico."[14] If the City of Holiness occupied a goodly portion of the Gulf of Mexico, it was indeed a mighty fortress of righteousness, a virtual nation unto itself. As such, Enoch's Zion symbolized the millennial kingdom of God which will roll forth until it fills the entire earth.

The loss of so extensive a portion of the earth did not go unnoticed. The word went forth: "Zion is fled!" (Moses 7:69).[15] Such incredible news must have set atingle the souls of the God-fearing who had been left behind. The more vicious elements of society would have received the news with unalloyed rejoicing. The City of Holiness stood as a living witness against sin and debauchery. Now that witness was no more. Little did men know that the flight of Zion spelled the doom of their dark world: six hundred four years later that world was swept from under heaven in a planet-wide deluge. But even as Enoch's Zion was taken from among men in a time of spiritual darkness, so will it return in a time of spiritual enlightenment to add its glory to that of the latter-day Zion of Joseph Smith.[16] Such was the assurance given to both Enoch and Noah by the Lord.

Post-Earth Ministry of Enoch

The Prophet Joseph Smith revealed the astonishing fact that Enoch and his people, having achieved a terrestrial condition, became ministers to peoples of like glory on other worlds of this eternity:

> Now this Enoch God reserved unto Himself, that he should not die at that time, and appointed unto him a ministry unto *terrestrial* bodies. . . .He is reserved also unto the Presidency of a dispensation....He is a ministering angel, to minister to those who shall be heirs of salvation. . . .Their place of habitation [as translated beings] is that of the terrestrial order, and a place prepared for such characters He held in reserve to be ministering angels unto many planets (TPJS, 170; see 191).

Since the dwelling place of translated beings is a terrestrial sphere, it may be that such men as the apostle John and the three Nephite disciples are associating with Enoch and his people in labors that extend to the "outmost parts of heaven."[17]

When shown the future sufferings of those who would perish in the Flood, Enoch's "heart swelled wide as eternity; and his bowels yearned; and all eternity shook" (Moses 7:41). His mission as a translated being took him into that eternity for which he had been spiritually prepared. Since he had established a terrestrial Zion on earth and had been appointed to minister to other terrestrial bodies, his mission seems to be centered in the progress of other Zion-like societies on other worlds. Indeed, he knew that God had always gathered such Zions unto himself out of those worlds.[18]

Ultimately, this multiplicity of Zions constitutes the one Zion of God even as the New Jerusalem is to be a complex of many Zion-like stakes or as the three Gods constitute the one God.[19] Unity has ever been the grand objective of the Lord. Christ declared: "If ye are not one ye are not mine" (D&C 38:27). This principle is universal; all men and all worlds must become one in Christ if he is to claim them as his own. Enoch's mission to "many planets" had this object in view. President John Taylor wrote:

> It would appear that the translated residents of Enoch's city are under the direction of Jesus, who is the Creator of worlds; and that He, holding the keys of the government of other worlds, could, in His administrations to them, select the translated people of Enoch's Zion, if He thought proper, to perform a mission to these various planets, and as death had not passed upon them, they could be prepared by Him and made use of through the medium of the Holy Priesthood to act as ambassadors, teachers, or messengers to those worlds over which Jesus holds the authority (*MA,* 76; *GK,* 103).

Melchizedek and His People

Melchizedek was a high priest after the order of the Son of God, a man of great righteousness and, therefore, of mighty faith.[20] This combination of virtue and authority endowed him with the priesthood power possessed by men who "were translated and taken up into heaven." The text implies that he, together with his people, eventually joined Enoch's Zion which had been translated some fifteen hundred years earlier:

> And his people wrought righteousness, and *obtained heaven*, and sought for the city of Enoch which God had before taken, separating it from the earth, having reserved it unto the latter days, or the end of the world...And this Melchizedek, having thus established righteousness, was called the king of heaven by his people, or, in other words, the King of peace (JST, Gen. 14:34, 36).[21]

THE GREAT DIVISION

Ninety years after the Flood, a man was born who is identified with one of the most remarkable events in geologic history. His name was Peleg. Genesis casually states, "in his days was the earth divided" (Genesis 10:25).[22] The Joseph Smith Translation reads: "And Peleg was a mighty man, *for* in his days was the earth divided" (JST, Gen. 10:16).[23] The use of the preposition "for" suggests that Peleg's mightiness had something to do with the miraculous division of the earth into its present continents—a miracle accomplished through the power of the Holy Priesthood which Peleg held. Such acts have been associated with righteous priesthood bearers from time immemorial.

> For God having sworn unto Enoch and unto his seed with an oath by himself; that every one being ordained after this

order and calling should have power, by faith, to break mountains, to divide the seas, to dry up waters, to turn them out of their course (JST, Gen.14:30).

Peleg, Enoch's descendant, was apparently endowed with just such power. Orson Pratt believed that whereas the Flood had undoubtedly "produced some changes on the surface of our globe," the division of "the one great antediluvian continent" into the various islands and continents did not occur at that time, but in the days of Peleg (JD, 18:317).[24] Regardless of the impact of the Fall and the Flood on the earth's primal land mass, its breakup into fragments constituting the present continents was completed during Peleg's lifetime—a period of two hundred thirty-nine years.

Many scholars assume Genesis 10:25 refers to a cultural or linguistic rather than a physical division of the earth. While this viewpoint is certainly more acceptable to the rational mind, it does not square with either ancient or modern revelation. For one thing, Genesis 10:32 states that the descendants of the three sons of Noah comprised *the nations* which were "divided in the earth after the flood." This is in contradistinction to the division thereafter *of the earth* which occurred during the lifetime of Peleg. Then too, Genesis 11 begins with the building of the tower of Babel, goes on to the confounding of man's universal language (with its consequent scattering of the peoples), and then pauses to provide a second abbreviated genealogy of the house of Shem ending with Abraham. This genealogy lists Peleg without restating the explanation given in Genesis 10:25 for his being so named.[25] The parenthetical placement of this modified genealogy of Shem's posterity—coming as it does after the account of the events associated with the tower of Babel—strongly suggests that the planet's division not only followed these events but was a direct consequence of them!

Pangaea and Continental Shift

Time is vindicating Moses. The theory that the earth's continents were once joined together has been confirmed within the last few decades. Francis Bacon is credited with advancing the theory four hundred years ago. It had a few advocates in the nineteenth century but its real impetus came in 1915 when Alfred Wegener, a German meteorologist and explorer, published a lengthy paper in which he asserted that the continents originally consisted of two great land masses which eventually broke up and began drifting apart. Like Bacon, Wegener, was disbelieved.

However, in the last few decades marine geologists have amassed a body of irrefutable evidence to the effect that all of the earth's land masses were once fused together into one supercontinent named Pangaea. In time, this "ur-continent" divided into what is now called Gondwanaland and Laurasia.[26] Then, about 230 million years ago[27] (due to further increasing pressures within the earth's interior) these two land masses began breaking up and drifting apart, thereby forming the continents as they exist today.[28] America supposedly joined Europe and Africa at a point about midway in the Atlantic Ocean, putting it about three thousand miles closer to Palestine than it is now.[29]

The suggestion that the formation of the major mountain ranges occurred in conjunction with the Flood and the division of the earth in the days of Peleg is theoretically supported by current geologic research. As before mentioned, it is reasoned that the crust of the earth (about ten miles deep), being composed of comparatively light materials, is "floating" on the heavier materials comprising the earth's mantle (about 2200 miles deep). The upper mantle has broken up into a number of relatively shallow sub-sections called "plates" that are about 60 miles thick and, in some instances, thousands of miles wide. The earth's internal pressures cause these plates to move about, sometimes slipping and sliding over and under one another. On occasion, in their random movements, the plates have caused their passengers—the continents—to collide, thereby causing them to buckle in the general vicinities of their points of contact. In this way, the Alps, the Andes, the Himalayas, etc., came into being.[30] That plate tectonics is unique to Earth and crucial not only to mountain building but also to the very existence of complex life is currently maintained by reputable astrobiologists.[31]

The breakup of the earth's original land mass, being causally identified with mountain-building, lends credence to the declaration of Moses in Genesis. However it occurred, that Earth has been divided is now the testimony of both scripture and science. Characteristically, the point at issue is not *what* happened, but *when* and *how* it happened. While it is desirable to receive scientific confirmation of scriptural claims, their ultimate validity is not determined by the position men take on them, whether pro or con. Modern scripture provides a second witness to Genesis 10:25 in an unmistakable allusion found in the Doctrine and Covenants. In a revelation given to the Prophet Joseph Smith in 1831, the Lord declared that when he comes again "the earth shall be like as it was in the days *before it was divided"* (D&C 133:24).

The Lord's wisdom in separating the Americas from Europe and Asia becomes apparent in the light of the prophetic history of Israel vis-a-vis the Gentiles. Doing so enabled him to apportion the earth among his spirit children according to his eternal purposes relative to the house of Jacob. Moses declared: "When the most High divided to the nations their inheritance, when *he separated the sons of Adam*, he set the bounds of the people according to the number of the children of Israel" (Deut. 32:8). Paul, mindful of Moses' words, told the Athenians that God "hath made of one blood all nations of men for to dwell on all the face of the earth, and hath determined the times before appointed, and the *bounds of their habitation"* (Acts 17:26).[32]

The Lord did not "set the bounds" of the nations until after the Flood and the division of the earth in the time of Peleg because the house of Israel did not come into being until a number of centuries after those spectacular events. This suggests that these "bounds" involved more than the mere assignment of different peoples to different portions of the earth. They were actually geologic demarcations designed to control and direct the flow of humanity from one portion of the globe to another. Israel's several lands of promise were scattered across the earth, being preserved behind natural barriers formed of mountains, deserts, jungles, and vast stretches of open sea.[33] This allowed the Shepherd of Israel to divide and/or isolate his flock as circumstance required. In time all shall realize that the earth was divided (as it was cursed) because of—and for the good of—mankind.

The division of this planet's original land mass is symbolic of the spiritual alienation which presently exists between God and mankind. The Almighty was much closer, in fact and in Spirit, to the antediluvians than he has been to Noah's posterity.[34] It is significant that the glorious reunion of the Creator, Jesus Christ, with the millennial earth will be accompanied by a like reunion of the earth's land masses. They will be "married" as they were in the beginning—before their "divorce" in the days of Peleg.[35]

THE LOST TRIBES OF ISRAEL

Israel's Beginnings

It was more than five hundred years after the Flood before the family of Israel came into being. Although Abraham is rightly regarded as the father of the Hebrews, the Israelitish wing of that clan was founded by his grandson, Jacob.[36] His four wives (Leah, Rachel, Bilhah and Zilpah) bore him twelve sons; their descendants comprise the "twelve tribes" of Israel.

The house of Israel was conceived in Mesopotamia, born in Canaan, and attained maturity in Egypt where it became a slave people with slave ways. Jehovah raised up Moses to offer them the truth that would make them free. They accepted only a portion of that truth and obtained only a portion of the physical and spiritual freedom they might have enjoyed. Subsequent to their exodus from Egypt and the passing of Moses and Joshua, they broke up into a loose-knit confederation of independent tribes held together as much by the constant threat of their alien neighbors as by any common commitment to the law of Jehovah. In time, however, their desire for the security and political status enjoyed by neighboring monarchies saw the period of the judges give way to the United Kingdom (c. 1020-922) under Saul, David, and Solomon.

Solomon's death sparked a civil revolt which ended in the division of the monarchy into the Southern kingdom of Judah and the Northern kingdom of Israel or Ephraim. The tribes of Benjamin and Simeon were generally absorbed into Judah, while the remaining tribes identified with Israel.[37] The two kingdoms existed side by side in an unstable peace for about two hundred years until, in a period of less than twenty years, the Northern kingdom was vanquished. By 738 it had become a mere vassal of Assyria. Six years later, Tiglath-pileser III (Pul) subdued much of Israel (Gilead, Galilee and the Plain of Sharon) and transported a number of the people to various localities in the empire.[38] Finally, in 721, a three-year siege of Samaria,[39] the capital city of Israel, ended in victory for Sargon II who "carried Israel away into Assyria, and placed them in Halah and in Habor by the river of Gozan, and in the cities of the Medes" (2 Kings 17:6).[40] Such was the origin of the so-called Lost Tribes.[41] They had been driven from their promised land because they had rejected Jehovah; they had betrayed their ancient covenants and descended to the grossest iniquities. Moses had warned Israel in clear and unmistakable terms that the breaking of its covenant with Jehovah would result in their expulsion from Canaan and their dispersion among the heathens:

> Ye shall be plucked from off the land whither thou goest to possess it. And the Lord shall scatter thee among all people, from the one end of the earth even unto the other (Deut. 28:63-64).[42]

Similar warnings were repeated by Israel's later prophets such as Amos:

> For, lo, I will command, and I will sift the house of Israel among all nations, like as corn [wheat] is sifted in a sieve, yet shall not the least grain fall upon the earth (Amos 9:9).

No true Israelite was to be lost forever. Commenting on the words of Isaiah, Nephi wrote, "the house of Israel, sooner or later, will be scattered upon all the face of the earth, and also among all nations."

> Behold, there are many who are already lost from the knowledge of those who are at Jerusalem. Yea, the more part of all the tribes have been led away; and they are scattered to and fro upon the isles of the sea; and whither they are none of us knoweth, save that we know that they have been led away (1 Ne. 22:3-4).[43]

The fulfillment of such prophecies began with the Northern Kingdom, the fate of which was declared by Ahijah more than two hundred years prior to the fall of Samaria:

> For the Lord shall smite Israel, as a reed is shaken in the water, and he shall root up Israel out of this good land, which he gave to their fathers, and shall scatter them beyond the river, because they have made their groves, provoking the Lord to anger (1 Kings 14:15).[44]

Although the phrase "beyond the river" is sometimes applied to the Jordan river system, the localities mentioned in 2 Kings 17 refer to the Euphrates—the river beyond which lay Halah and Harbor and the cities of the Medes.[45] Indeed, they were the very areas from which Sargon II selected those who were to replace the Israelites in Samaria.

The Samaritans

Applying the principle of divide and conquer, Sargon II made it a policy to exile a considerable portion of each captive people to distant regions of the empire where they could be assimilated into other minorities. This policy was followed with the Israelites: "People from the lands which I had conquered I settled there." Many thousands of Israelites were uprooted and exiled from their promised land; their places were taken by idolaters:

> And the king of Assyria brought men from Babylon, and from Cuthah,[46] and from Ava, and from Hamath, and from Sepharvaim, and placed them in the cities of Samaria instead of the children of Israel: and they possessed Samaria, and dwelt in the cities thereof (2 Kings 17:24).

In time, many of those Israelites left behind in northern Palestine intermarried with their heathen neighbors, thereby producing an ethnically and religiously mixed people—the Samaritans. Regarded as mongrelized inferiors, the Samaritans were despised by the Jews even in the days of Jesus. Not all tribal members remained in the north; some moved to Judah even before the fall of the Northern Kingdom and, in time, others followed.[47] This was easily done since free movement from place to place had never been challenged. Indeed, there was considerable travel and commerce both within and between nations of the ancient world. The number of Israelites who left their homeland and immigrated to other lands—whether voluntarily or otherwise—was quite large. Thus a process of dispersion spanning many centuries resulted in the sprinkling of the blood of all twelve tribes throughout the earth. The sheep of the Holy One of Israel were scattered on a thousand hills.

A Remnant Led Away

Scripture indicates that the scattering of Israel was to be of a dual character. On the one hand it involved the assimilation of some Israelites into various ethnic groups to such an extent that their Israelitish origins were lost and they became, to all intents and purposes, Gentiles.[48] On the other hand it involved the preservation of selected remnants from the different tribes by means of geographic and/or cultural isolation.[49] Such was the case with ancient Israel; Jehovah reserved a remnant of those taken captive to Assyria for himself— "even the tribes which have been lost, which the Father hath led away out of Jerusalem" (3 Ne. 21:26).[50] The Savior seems to distinguish between those Israelites *carried* away by Sargon II to the *known* land of Assyria and those Israelites *led* away by the Father to an *unknown* land described by the Savior as being "neither of the land of Jerusalem, neither in any parts of that land round about whither I have been to minister" (3 Ne.16:1).[51] That land was unknown to everyone but God— "for they are not lost unto the Father, for he knoweth whither he hath taken them" (3 Ne. 17:4). Thus, the Lost Tribes of prophecy were drawn from the main body of exiles not long after their forced emigration from the promised land. They are the choice remnant of that remnant of the Northern Kingdom of Israel which faded from religious history almost three thousand years ago.

Although there is no extant scriptural record of the exodus of the Ten Tribes from Assyria, the apocryphal book, 2 Esdras (Ezra), provides an account compatible with all that has been revealed on the subject:

> And whereas thou sawest that he [God] gathered another peaceable multitude unto him; those are the ten tribes, which were carried away prisoners out of their own land in the time of Osea [Hosea] the king, whom Salmanasar the king of Assyria led away captive, and he carried them over the waters, and so came they into another land. But they took this counsel among themselves, that they would leave the multitude of the heathen, and go forth into a further country, where never mankind dwelt, that they might there keep their statutes, which they never kept in their own land. And they entered into Euphrates by the narrow passages of the river. For the Most High then shewed signs for them, and held still the flood, till they were passed over. For through that country there was a great way to go, namely, of a year and a half: and the same region is called Arsareth (2 Esdras 13:39-45).

Commenting on this passage, George Reynolds suggested that the only place the tribes could go and be free of all contaminating influences lay to the north "toward the polar star." He believed that they crossed the river Euphrates at a gorge far to the north "so narrow that it is bridged at the top." Reynolds concluded: "How accurately this portion of the river answers to the description of Esdras of the 'narrows' where the Israelites crossed!"[52] In his imaginative description of the further travels of Israel "toward the polar star," Reynolds wrote:

> . . .inasmuch as they had turned to the Lord and were seeking a new home wherein they could the better serve Him, they were doubtless guided by inspired leaders. . . .But what

must have been their sensations when they came in view of the limitless Arctic Ocean. . . .The prospect must have been appalling to the bravest heart not sustained by the strongest and most undeviating faith in the promises of Jehovah. . . .No wonder if some turned aside, declared they would go no further and gradually wandered back through northern Europe to more congenial climes (Reynolds, 31-33).[53]

It is commonly understood that a considerable number drifted into Northern Europe and Britain where they eventually merged with the various Teutonic tribes. The fact that the ancestries of most early members of the LDS Church are traceable to that region and that they have been declared to be descendants of Joseph's son, Ephraim, is considered highly supportive of this belief.[54]

That there were those of the Northern Kingdom who repented of their past follies and sought to isolate themselves from the heathen world that had been so instrumental in bringing about their downfall is consistent with the ways of the Lord.[55] He has endeavored to remove his people from among the cultures of the world on several occasions. Enoch isolated his people in his City of Holiness. The Jaredites were separated from the people of Babel and brought to a virgin America. Moses and Joshua were instructed to utterly destroy the Canaanites from among them. A remnant of Joseph was led away from Palestine so that they could be isolated from the corrupting influences of the Jews in a land where God could "raise up unto me a righteous branch from the fruit of the loins of Joseph." And a remnant of that people was led away by Mosiah I prior to the first destruction of the Nephite nation. It is a common scriptural motif.

Jehovah's Other Sheep

There is scriptural evidence that the Lost Tribes did repent and become faithful to the law of Moses and even worthy of receiving the higher gospel law. It is provided in a prophecy of Zenos. He predicted that in due time the Lord God would manifest himself to all the house of Israel and that he would visit *"some with his voice, because of their righteousness"* (1 Ne. 19:11). The Savior fulfilled this prophecy by personally teaching those of the house of Joseph in America who were spared in the destructions associated with his crucifixion because they were "more righteous" than those who perished. He explained to them that he had "other sheep" which were not located either in America or in Palestine who had never heard his voice or received his personal ministrations. By command of the Father, he was going to visit them so that they might be "numbered among my sheep" (3 Ne. 16:1-3). That he was referring specifically to the Lost Tribes is made clear by his announcement to the Nephites:

> But now I go unto the Father, and also to show myself unto the lost tribes of Israel, for they are not lost unto the Father, for he knoweth whither he hath taken them (3 Ne. 17:4; see 21:26-29).

So at the time the resurrected God of Israel ministered to the house of Joseph in America, he also ministered to the Lost Tribes. They, too, beheld his face and heard his voice because, as Zenos had prophesied, "of their righteousness."[56]

Although the mortal Messiah ministered to the gathered sheep of the Jewish fold, only a select few heard his voice and beheld his glory following his resurrection. Nor did he visit any lost sheep among the Gentiles. The fold of Joseph in America and the fold of the Lost Tribes in their unknown location were each assembled apart from the nations of the earth. While the servants of the Lord are obliged to cover the world as fishers and hunters of men, the Shepherd of Israel has not and will not do so. His personal mission is always limited to a physically gathered people.

Parenthetically, there is no reliable evidence whatsoever that the Savior visited any of the scattered blood of Israel in the nations of Europe in 34, or at any other time subsequent thereto. The identification of Jesus Christ with such pagan deities as Odin, the supreme god of Norse mythology, is quite unsupportable. Nor is there any valid basis for the claim that Christ actually visited Britain or any other European country following his resurrection.[57] Their inhabitants were Gentiles who were ineligible for a personal visitation from the resurrected Christ. For he had told the Nephites, "the Gentiles should *not at any time hear my voice—that I should not manifest myself unto them save it were by the Holy Ghost*" (3 Ne. 15:23). Whatever the presence of the blood of Israel may be among the Gentile peoples of Europe, it does not solve the mystery of the whereabouts of those "other sheep" whom the Shepherd of Israel visited following his initial appearance to his fold in America.

VIEWS ON LOCATION OF LOST TRIBES

The present whereabouts of the Lost Tribes has intrigued students of the Bible for centuries. The question has produced a number of answers ranging from the rationally plausible to the seemingly impossible. For Latter-day Saints their location remains a matter of opinion; no official statement from the First Presidency on the subject has been ever made. The traditional view has been that one portion of these tribes was dispersed among the nations while another portion remains hidden away. But for this to be the case, a miracle must be introduced into the matter. For how can there be a civilization numbering in the hundreds of thousands, if not millions, somewhere on the earth unknown to the rest of mankind? Granted, in the past small groups of primitives have been discovered from time to time in some remote area of the globe, but that is a far cry from the sort of thing envisioned for the Lost Tribes. Where *in the world* are they? Herewith, the main theories.

Polar Regions

In his previously cited work, George Reynolds took the position that the location of the Lost Tribes in the "frozen regions of the north" was the belief of the Latter-day Saints.[58] This was generally accepted in the 19th century when the polar regions were largely unexplored, but in the main it has not been publicly advocated by Church authorities for well over half a century. Jeremiah prophesied the Lost Tribes would come from (or out of) the "land of the north" or "the north country."[59] He is supported by Moroni who, in summarizing the teachings of the more ancient prophet Ether, wrote that the inhabitants of old Jerusalem will be "gathered in from the four quarters of the earth, *and* from the north countries" (Ether 13:11). Two modern revelations received by Joseph Smith also place the Lost Tribes in "the north countries" and declare they will be led "from the land of the north" (see D&C 110:11; 133:26).

The phrase "land of the north" has been generally understood to refer to the unpopulated regions of the globe beyond the Arctic Circle. Consequently most of those who adhered to the unified body concept argued that the Lost Tribes were hidden away somewhere in the northern polar regions. However, not one of the passages bearing on the subject explicitly supports such an interpretation. This did not prevent the notion from gaining a foothold both in and out of the Church among some commentators. Several theories involving that general locale have been advanced.

Benjamin F. Johnson asked the Prophet Joseph Smith "where the nine and a half tribes of Israel were." Johnson says he was told, "they are in the north pole in a concave just the shape of that kettle. And John the Revelator is with them, preparing them for their return."⁶⁰ This theory was ridiculed by one reader in the old *Deseret Weekly:*

> Some think the earth hollow and that at the northern end of the earth there is a great hole. They fancy that the earth is inhabited inside with a race of people, said by some to be what is called the ten tribes, as the statement is made that they journeyed to the north for many days and it seems impossible to many to account for them on the land that they now live on. . . .I cannot believe that they are in any such a locality.⁶¹

The second theory attributed to Joseph Smith is based upon the claimed testimony of his body guard, Philo Dibble.⁶² According to the hearsay account, a small sphere lies above and beyond each of the poles on a line with the earth's inclined axis. These two spheres constitute, as it were, the wings of the earth, with the Lost Tribes supposedly located on the north "wing."⁶³ A "narrow neck of land" connects each of the spheres to the earth.⁶⁴ Both Johnson and Dibble were men of recognized integrity, however, their accounts appear to be garbled versions of the Prophet's remarks.

Among the earliest LDS leaders known to have advocated a polar setting for the Lost Tribes were W. W. Phelps and Orson Pratt. After theorizing that there might be more than three and a half *billion* souls living in unknown parts of the earth, W.W. Phelps continued:

> Let no man marvel at this statement, because there may be a continent at the north pole, of more than 1300 square miles, containing thousands of millions of Israelites.

He further suggested that they were the branches referred to in the allegory of Zenos which were planted in the nethermost parts of the earth which brought forth much fruit—since "no man that pretends to have pure religion, can find 'much fruit' among the Gentiles, or heathen of this generation."[65]

Orson Pratt shared Phelps' expansive notion of an arctic civilization "around the pole" not unlike the Shangri-la of Hilton's *Lost Horizon*. He described it as being an unknown region "seven or eight hundred miles in diameter" encircled by "great mountain ranges" and characterized by "deep and extensive valleys" having comparatively mild temperatures.[66] Others, while denying that a polar location for the tribes was either scriptural or doctrinal, have expressed the belief that they could still be hidden from men somewhere in the Arctic. In 1919, commenting on Doctrine and Covenants 133:27, Smith and Sjodahl wrote that Peary's 1909 exploration of the North Pole area did not prove that the Lost Tribes were not hidden away in the Arctic because "there is a great deal of country in the north that no man, to our knowledge, has visited. . . .[the Lord can] keep them hidden. . .until it is time for them to be revealed."[67]

However the more we learn about this planet the less tenable are the theories relative to the North Pole region. The extensive mapping and settlement of the Arctic, together with the advanced state of electronic surveillance renders the idea of a northern hiding place for the Lost Tribes increasingly unlikely. Satellites equipped with cameras capable of accurately photographing virtually any and all objects on the ground below are constantly sweeping across much of the planet. Thanks to such technology, probing eyes are everywhere; there is simply no place to hide even a small settlement much less an entire civilization. From a rational standpoint it is most improbable that they are to be presently located anywhere in the northern reaches of this planet. Orson F. Whitney defended their existence, not their location:

The fact that Arctic explorers have found no such people at the North Pole—where some theorists have persisted in placing them—does not prove that the "Ten Tribes" have lost their identity. It was *tradition, not revelation,* that located them at the North Pole. . . .Those tribes could still be intact, and yet much of their blood be found among the northern nations (*SNT,* 174, note "s").

Scattered Among the Nations

The most widely-held position among Bible scholars is that the tribes are not lost to the world but lost in the world—absorbed into the milieu of the nations.[68] B.H. Roberts of the Seventy was the first Church authority to publicly subscribe to this view.

> I believe, for myself, that within the known regions of the earth, where the children of men are located, it is quite possible for God to fulfill all of his predictions in relation to the return of Israel. . . .[They would not] be lost to the knowledge of God, though nowlost to men. And as it was possible to lose these tribes of Israel among the nations of the earth, so is it possible for God to recover them from their scattered condition from among these nations, with a display of the divine power.[69]

In *Mormon Doctrine* (1958, 1966), Bruce R. McConkie wrote of the Lost Tribes as an integral group visited by the risen Christ who would return with "their prophets and their scriptures."[70] However, in 1985 he wrote:

> . . . .we know only that the Lost Tribes are scattered in all the nations of the earth, and that when they are gathered, it will be on the same basis and in the same way that any converts are made. . . .the Ten Tribes are to come back like anyone else: by accepting the Book of Mormon and believing the restored gospel. . . .the whole house of Israel the Ten Tribes

included, will be gathered one by one as their hearts are touched by the Spirit of Christ.[71]

A Single Body

Parley P. Pratt distinguished between the condition of the Jews and that of the Lost Tribes. The Jews, said he, were said to be *dispersed* because they had been "scattered among the nations." In contrast, the Lost Tribes are "*outcasts* because they are cast out from the knowledge of the nations into a land by themselves" (VW, 29). This distinction is implied in the keys bestowed upon Joseph Smith and Oliver Cowdery by Moses for "the gathering of Israel. . .*and* the leading of the ten tribes from the land of the north" (D&C 110:11). And such is the implication of Joseph Smith's declaration in the tenth Article of Faith: "We believe in the literal gathering of Israel *and* in the restoration of the Ten Tribes." The Prophet was not being redundant; he is writing of two different events. In refuting the belief that Israel exists "only in a scattered condition," Orson F. Whitney wrote:

> If this be true, and those tribes were not intact at the time Joseph and Oliver received the keys of the gathering, why did they make so pointed a reference to "the leading of the ten tribes from the land of the north?" This, too, after a general allusion to "the gathering of Israel from the four parts of the earth." What need to particularize as to the Ten Tribes, if they were no longer a distinct people? And why do our Articles of Faith give those tribes a special mention? (*SNT*,174).

Joseph Fielding Smith quoted this statement with approval and added:

> That they are intact we must believe, else how shall the scriptures be fulfilled? There are too many prophecies

concerning them and their return in a body, for us to ignore this fact (*WTP*,130).

James E. Talmage took a similar position:

. . .while many of those belonging to the Ten Tribes were diffused among the nations, a sufficient number to justify the retention of the original name were led away as a body and are now in existence in some place where the Lord has hidden them (*AF*, 340).

In 1916 Talmage reaffirmed this earlier belief:

There are those who would juggle with the predictions of the Lord's prophets. . . .I have found elders in Israel who would tell me that the predictions relating to the Lost Tribes are to be explained in this figurative manner— that the gathering of those tribes is already well advanced and that there is no hiding place whereto God has led them, from which they shall come forth, led by their prophets to receive their blessing here at the hands of gathered Ephraim, the gathered portions that have been scattered among the nations. . . .The tribes shall come; they are not lost unto the Lord; they shall come forth as hath been predicted; and I say unto you there are those now living— aye, some here present— who shall live to read the records of the Lost Tribes of Israel (CR, Oct, 1916, 76).

Beyond the Earth

Since the Lost Tribes have not been found among the inhabited regions of our globe, and since they are referred to as being in or coming from the north, it is to the north that we must look. But, as previously noted, the Arctic region has been well-charted and subjected to ever-increasing development. For a population numbering in the many thousands, if not millions, to be hidden away there, either on or within the earth, would require an unprecedented and uncharacteristic miracle on the Lord's part.[72] Although Scripture indicates the Lost Tribes will come *from* the north, nowhere is it written that they actually dwell *in* the northern reaches of this planet.

Moses assured Israel that although they would be scattered, in a day of repentance they would be "gathered from all the nations." He added: "If any of thine be drive out unto the *outmost parts of heaven,* from thence will the Lord thy God gather thee, and from thence will he fetch thee" (Deut. 30:40).[73] Was this mere hyperbole? Was Moses simply underscoring the certainty of the Lord's promise? Or was he alluding to another segment of Israel that would not be gathered out of the nations but brought back from the "outmost parts of heaven"?[74]

Did a remnant of Israel follow the precedent of Enoch? Were they spirited away from the earth? According to the testimonies of a number of his associates, the Prophet Joseph Smith did reveal an extra-terrestrial location for the Lost Tribes.[75] Charles L. Walker recorded several such testimonies in his private journal. In his 18 October 1880 entry, he records Addison Everett's remarks:

> Br. Everett said that he heard Joseph say that the earth had been divided and parts taken away, but the time would Come when all would be restored and the earth would revolve in its original orbit next to Kolob and would be second in size to it.[76]

On Sunday, 6 March 1881 Charles Walker heard Jacob Gates recount the Prophet's remarks in 1838 on the return of the Lost Tribes:

> Said he heard Joseph Smith say when he was at Bishop Partridges House in Far West, Missouri, concerning the ten Lost Tribes, They are hid from us by land and air. Said Bishop [Edward] Partridge, I guess they are by land and water, in a doubting manner as if Joseph did not know what

he was talking about. Yes, said Joseph, by land and air; they are hid from us in such a manner and at such an angle that the Astronomers cannot get their telescopes to bear on them from this Earth (Walker, 2:539).

Daniel Allen's testimony also confirmed the Prophet's belief in an extra-terrestrial location for the Lost Tribes:

I heard Joseph the Prophet say that he had seen John the Revelator and had a long conversation with him, who told him that he John was their leader, Prophet, Priest and King, and said that he was preparing that people to return and further said there is a mighty host of us. And Joseph further said that men might hunt for them but they could not find them for they were upon *a portion of this planet that had been broken off* and which was taken away and the sea rushed in between Europe and America, and that when that piece returns there would be a great shake; the sea would then move to the north where it belonged in the morning of creation.[77]

Key elements in this statement are in harmony with other accounts of the Prophet's teaching. Bathsheba W. Smith reported him saying:

Peradventure, the Ten Tribes were not on this globe, but a portion of this earth cleaved off with them, went flying into space, and when the earth reels to and fro like a drunken man, and the stars from heaven fall, it would join on again.[78]

After praising Joseph Smith's ability to clarify and expand the meaning of scripture, Wandle Mace cited an occasion when he had reached an impasse as to the whereabouts of the Lost Tribes since—if they had multiplied proportionately to the Jews— "where is there habitable earth for this vast amount of people who are hidden from the rest of mankind?" His question was answered by the Prophet in a Sunday morning discourse on the restitution of all things:

In the course of his remarks he spoke of the Earth being divided at various times he said "When Enoch and his City were taken away, a portion of the Earth was taken, and it would be again restored. Also in the days of Peleg, the earth was divided. . . ." He then referred to the Ten Tribes saying, "You know a long time ago in the days of Shalmanezer King of Assyria, the Ten Tribes was taken away, and have never been heard of since." He said, "The Earth will be restored as at the beginning, and the last taken away, will be the first to return, for the last shall be first, and the first shall be last in all things". . . .These remarks satisfied me, it was no longer necessary to hunt the place on this earth where the Ten Tribes were so long hidden, for the earth was divided and taken away, and will be the first to return, as it was the last taken away (Wandle Mace, 35).

Parley P. Pratt expressed the same basic idea: fragments of the earth have been broken off— "Some in the days of Enoch, some perhaps in the days of Peleg, some with the ten tribes, and some at the crucifixion of the Messiah."[79]

Among other notable persons who gave similar testimony was Eliza Roxey Snow. One stanza of her poem, "Address to Earth," reflects her understanding of the subject:

> And when the Lord saw fit to hide
>   The "ten lost tribes" away,
> Thou, earth, wast sever'd to provide
>   The orb on which they stay.

Since this lyric was included in Church hymnals for some fifty years, it would appear that its doctrinal acceptability was acknowledged during the administrations of Brigham Young, John Taylor, Wilford Woodruff, and Lorenzo Snow. There is good reason to believe that Eliza R. Snow obtained her views on the Lost Tribes—as well as other doctrines—from Joseph Smith. In a signed statement, Homer M. Brown described an experience had by his grandfather, Benjamin Brown:

> Brother Brown, will you give us some light and explanation of the 5th verse on page 386 of the Hymn Book which speaks of the Ten Tribes of Israel, or the part of this earth which formed another planet, according to the Hymn of Eliza R. Snow?
>
> "Yes, sir, I think I can answer your question. Sister Eliza R. Snow, in visiting my grandparents, was asked by my grandmother: 'Eliza, where did you get your ideas about the Ten Lost Tribes being taken away as you explain it in your wonderful hymn?"
>
> She answered as follows: "Why, my husband (The Prophet Joseph) told me about it."[80]

In confirmation of Eliza R. Snow's testimony, Patriarch Brown then told of a conversation between his grandfather and the Prophet concerning the Ten Tribes. Responding to the question of their location, Joseph Smith is said to have taken Benjamin Brown and his wife outside and, pointing to the north star, asked them: "Now do you discern a little twinkle to the right and below the Polar Star, which we would judge to be about the distance of 20 feet from here? They answered in the affirmative and then returned to the house" (Smith, 211-13).

This is a much-ridiculed story, but Brown's testimony is corroborated in two entries in the previously quoted journal of Charles L. Walker, a friend of Eliza R. Snow. He records that in a prayer meeting on 11 February 1881, a "Sister Green" told of hearing Eliza R. Snow "speak of the 9 and 1/2 Lost Tribes being on an orb and would eventually come back to the[i]r former place and we should know when they came by certain signs etc., etc." (Walker, 2: 532). These remarks may have prompted Walker's visit to Eliza R. Snow a month later during which the division of the earth was discussed:

> At night paid Sister Eliza R. Snow a Short Visit and had some conversation with her on the Dividing of the Earth. She told me that she heard the Prophet Joseph say that when the

10 tribes were taken away the Lord cut the Earth in two, Joseph striking his left hand in the center with the edge of his right to illustrate the idea, and that they (the 10 tribes) were on an orb or planet by themselves, and when they returned with the portion of this Earth that was taken away with them, the coming together of these 2 bodies or orbs would cause a shock and make the "Earth reel to and fro like a drunken Man." She also stated that he said the Earth was not [now] ninety times smaller now than when first created or organized (Walker, 2:540).[81]

There is no valid basis for questioning either the Walker entry or Eliza R. Snow's essential memory. Wilford Woodruff's journal entry for 8 September 1867 records Brigham Young's attribution of the extra-terrestrial concept to Joseph Smith:

We had social conversation in the evening. President Young said he heard Joseph Smith say that the Ten Tribes of Israel were on a portion of land separated from this earth.[82]

In his biography of Wilford Woodruff, Matthias Cowley summarizes the foregoing journal entry as follows:

The leaders on their return from Provo made a visit to Logan. Here, President Young is quoted as saying that the ten tribes of Israel are on a portion of the earth—a portion separated from the main land (Cowley, 448).

We might dismiss one or two such statements as being mere second and third hand hearsay, but what are we to do with a *pattern* of them from prominent, responsible individuals who were closely associated with the Prophet for a number of years? Although some elements in their respective testimonies are at variance, all agree on the essential point. But, again, it is not the details which are significant, it is the *very existence of multiple accounts* testifying to the same basic proposition. The combined testimonies of such close associates of the Prophet as

Brigham Young, Wilford Woodruff, Parley P. Pratt, Eliza R. Snow, Bathsheba W. Smith, and Wandle Mace should not be lightly dismissed. Details aside, these witnesses are united in testifying that Joseph Smith consistently and repeatedly located the Lost Tribes beyond the Earth.

Conclusion

If the present whereabouts of the Lost Tribes were to be made known to anyone, it would be the Lord's prophet. But so far as the public record is concerned, none of Joseph Smith's successors have claimed personal revelation on the matter. Never once did the Prophet Joseph suggest the Lost Tribes were in a scattered condition. The available evidence is that he repeatedly referred to their extra-terrestrial location. The flight of Enoch and his city to a terrestrial sphere establishes a precedent for the Lost Tribes. For if they are not hidden on the Earth, they must be hidden in the heavens—otherwise, as Elder Joseph Fielding Smith asked, "How shall the scriptures be fulfilled?"

NOTES

1. For statements by Church authorities on the earth's original land mass see *Voice of Warning*, 85; *Man, His Origin and Destiny*, 381-382; *Mormon Doctrine*, 289.
2. One evidence of this likelihood is that dinosaur fossils are found on all continents.
3. Enoch lived circa 3378-2948 B.C.—four hundred thirty years before being translated; see D&C 107:48-49.
4. See D&C 107:48.
5. See Moses 6:26-36; D&C 76:67; JD, 16:48; 21:242; 26:24.
6. See Moses 7:19. Adam's blessing upon Enoch was probably in connection with the establishment of his city; see D&C 107:48.
7. See D&C 38:4; 45:12.
8. See Moses 7:21; JD, 3:320; 16:49.
9. Being translated, the resurrection of Enoch and his people is apparently still future. Joseph Smith explained, "translation obtains deliverance from the tortures and sufferings of the body, but their existence will prolong as to the labors and toils of the ministry, before they can enter into so great a rest and glory" (TPJS, 171).
10. See D&C 76:67.
11. Autobiography of Wandle Mace, (ms. copy), Brigham Young University, H. B. Lee Library, 35.
12. This was also the belief of Heber C. Kimball (JD, 8:107), John Taylor (JD, 21:157; 26:34, 89- 90), Wilford Woodruff (JD,11:242), and Orson Pratt who said that Enoch's people "having learned the doctrine of translation, were caught up into the heavens, the whole city, the people and their habitations" (JD, 17:147; see 17: 8:51; 15:263; 16:49-50).
13. Joseph Young, Sr., *History of the Organization of the Seventies* [Salt Lake City: Deseret News, 1878], 11. Since Enoch and his people are with Christ who dwells in "the regions which are not known," it would appear unlikely for them to be found in this solar system; see D&C 45:11-12; 133:46.
14. Joseph Young, 12.
15. The verb, "fled," is noteworthy. Men did not say that the city had burned, been destroyed, or sunk into the earth or sea. They said that it had "fled," that it had actually disappeared or taken flight. There may have been eye-witnesses to the event (see JD, 2:212; 21:157; 26:34).
17. See TPJS, 85; Deut. 30:4.
18. See Moses 7:31.
19. See Moses 7:64; D&C 20:28.

20 Melchizedek bestowed the Holy Priesthood upon Abraham and blessed him with all of the riches of eternity (see JST Gen. 14:40; D&C 84:14).
21 A precedent for such translations to Enoch's city was established prior to the Flood when worthy saints "were caught up by the powers of heaven into Zion" (Moses 7:27).
22 In Hebrew, Peleg means "division." Referring to Adam's possible dwelling place after his fall, Orson Pratt said: "It might have been upon what we now term the great eastern hemisphere, for in those days the eastern and western hemispheres were one, and were not divided asunder till the days of Peleg" (JD, 16:48). However, according to Joseph Smith, Adam lived in an area of what is now Missouri.
23 Peleg lived from 1746 to 1985 after the Fall, or between circa 2254 and 2015. These dates are based on the genealogies found in Genesis and Moses. Abraham may have been alive when the earth was divided since he was about forty-eight years old when Peleg died.
24 See JD, 8:195; 17:187; *Voice of Warning,* 87. Bruce R. McConkie wrote, "the so-called geological changes in the earth's surface, which according to geological theories took place over ages of time, in reality occurred in a matter of a few short weeks incident to the universal deluge" (*Mormon Doctrine,* 289).
25 See Gen. 11:16.
26 Gondwanaland (the Southern Hemisphere) formed Antarctica, Australia, South America, India, and Africa. Laurasia (the Northern Hemisphere) formed North America, Greenland, Europe, and most of Asia.
27 Geologists theorize that Pangaea was formed about 220 million years ago out of existing "continental masses." The present continents were supposedly formed out of Pangaea in the last two hundred million years. This figure is based on the present distance between North America and Europe and the belief that the Atlantic Ocean floor is now expanding at the rate of about one inch per year while the Pacific Ocean floor is contracting. It is a matter of simple arithmetic. Geologists disagree on the cause of the breakup of Pangaea. Melvin Cook, an LDS scientist, argues that the cause was not "drift" but *"shift."* That is, it involved a relatively sudden, rapid breakup of Pangaea, not a slow, "natural" drift (see Cook, *Scientific Prehistory,* chaps. 6-8).
28 See D&C 88:45.
29 A number of models have been suggested for Pangaea, but its actual size and shape are unknown.
30 The presence of marine fossils on very high mountains is believed to be an evidence that the Himalayas were formed from materials forced up from the ocean floor when the plate carrying India and Australia ground into the Eurasian landmass.

31 See Ward and Brownlee, *Rare Earth.*
32 See, *The Way To Perfection,* 46-47.
33 See 1 Ne. 17:38; 2 Ne. 24:2; Isa. 24:1-2.
34 Brigham Young taught: "The things that pertain to God and to heaven were as familiar among mankind, in the first ages of their existence on the earth, as these mountains are to our mountain boys, as our gardens are to our wives and children, or as the road to the Western Ocean is to the experienced traveller. From this source mankind have received their religious traditions" (JD 9:148).
35 See Isa. 62:1-4; HC, 1:275.
36 The name, Israel (ruling with God), was bestowed upon Jacob when he returned to Canaan after his twenty-year sojourn in Haran (see Gen. 32:24-28; 35:10).
37 See 1 Kings 11:29-36.
38 See 2 Kings 15:29.
39 The siege was prompted by Israel's rebellion against Assyria following an ill-advised treaty with Egypt which Isaiah warned against.
40 The annals of Sargon II state that 27,290 Israelites were taken captive. However, this figure is probably limited to those more prominent citizens taken in the fall of Samaria; the actual number exiled to Assyria through the years was undoubtedly much larger.
41 Neither in the Bible nor the Doctrine and Covenants are these tribes referred to as "lost." However, the term does appear in the Book of Mormon. See 2 Ne. 29:13; 3 Ne. 17:4; 21:26. The designation "ten" is found only in 1 Kings 11:31-35 and D&C 110:11.)
42 See Deut. 4:25-28; 28:25-37; Josh. 23:15.
43 See Isa. 49. The allegory of Zenos (Jacob 5) deals with the varied scatterings of Israel.
44 See 1 Kings11:31-36. The "groves" were the places of idolatrous worship.
45 H. Agrippa warned the Jews—should they attempt to revolt— not to look for assistance from their compatriots "beyond the Euphrates" in Parthia (see F. Josephus, *Wars of the Jews,* Book II, chap. 16:4). The apocryphal writing, 2 Esdras, says the Lost Tribes were carried "over the waters" into another land from whence they eventually crossed the Euphrates in search of "a further country" (see 2 Esdras 13:39, 42).
46 Josephus, the first century Jewish historian, wrote that the Assyrians replaced the Israelitish exiles with idol worshippers from Cuthah in Persia (see F. Josephus, *Antiquities of the Jews,* Book IX, chap. 14).
47 See 1 Chr. 11, 15. Note that although Lehi lived in Jerusalem, he was from the tribe of Manasseh. Ishmael, a relative and fellow resident, was from the tribe of Ephraim (see JD, 23:184; 1 Chr. 9:3; Luke 2:36).
48 This process had been underway long before the fall of the kingdom of

Israel and has continued to the present day. Note the marriage of Judah (Gen. 38) and the fact that a "mixed-multitude" accompanied Israel out of Egypt.(Ex. 12:38. See Num. 11:4; Lev. 24:10; Hosea 7:8 and Neh.13:3). The Restored Church is described in Gentile terms even though its members are primarily descendants of Jacob (see 1 Ne. 10:14; 13:35-40; 15:13; 21:22; 22:8; 2 Ne. 30:2,3; 3 Ne.16:6-7; Ether 12:22; Morm. 5:10,15; D&C 14:10; 109:60).

49 An example of cultural insulation are the Jews who have retained their fundamental ethnic identity in spite of their worldwide dispersion. Examples of geographic isolation are the people of Enoch, the Jaredites, the Lehites, and the Mulekites.

50 The redemption of the Lost Tribes is but one phase of the "work of the Father" in the latter days (see 3 Ne. 15:15-17; 16:1-3; 17:4).

51 The whereabouts of at least some members of the Lost Tribes was supposedly known to the ancient world, as is attested by Flavius Josephus (Wars, Book II ch. 16, par.4) and the fifth century church scholar, Jerome, who wrote that the Lost Tribes dwelt in the land of the Medes where they were still subject to Persian rule (*Commentaries,* vi, 7, 30).

52 George Reynolds, *Are We of Israel?*, 27- 28. This work is based in part on Reynolds' articles appearing much earlier in the *Juvenile Instructor* and was copyrighted by Joseph F. Smith for the Church's Deseret Sunday School Union.

53 See also *Juvenile Instructor,* 18 (15 January 1883), 28.

54 Brigham Young taught that the Anglo-Saxons were the descendants of Ephraim (see JD, 10:188).

55 Orson Pratt wrote: "They must have repented of their sins or God would not have miraculously divided the river for them to pass over" (*Millennial Star,* 29 (30 Mar. 1867), 200; see 3 Ne. 16:1-3).

56 Joseph Fielding Smith said, "These Lost Tribes were in a body somewhere when the Savior visited the Nephites on this continent. We believe he went to them and established his Church among them with an organization similar to that given to the Nephites. They had their prophets and kept a record" (*The Way To Perfection,* 131).

57 In speaking affirmatively of the British-Israel movement, Anthony W. Ivins (of the First Presidency), expressed his belief that Israel was in the British Isles "where we have always known them to be" (CR, 3 Oct. 1926, 16-18). However, Elder Mark E. Petersen, in rejecting the British-Israel doctrine, added: "I do not believe we should accept the current views that the lost ten tribes have been found in the northern nations of Europe, or that they have been named, indexed, and classified. I do not believe that we can accept the peculiar notion that the mythical Odin of the North was in reality the Savior of the world performing his work among the northern nations of Europe or the ten

tribes" (CR, 5 April 1953, 83). Bruce R. McConkie also rejected the "Bristish-Israel" movement and its identification with the Ten Tribes (see *A New Witness for the Articles of Faith*, 517).
58 See Reynolds, 8, 10, 23, 24, 33.
59 See Jer. 3:18; 16:15; 23:8; 31:8.
60 Benjamin F. Johnson, 93.
61 W.J.R, "The North Pole," *The Deseret Weekly,* (20 June 1896), 20-21.
62 It is a third person account insofar as Philo Dibble is concerned. His son Sidney signed a notarized affidavit in 1906 to the effect that a drawing shown him by M.W. Dalton was a reasonably accurate facsimile of a diagram shown him by his father. He further attested that his father told him that Joseph Smith had given him the diagram in about 1842. However, Sidney did not attest to any explanation of the diagram (see Dalton, 86).The notion of a sphere joined to the earth at or near its northern axis by a narrow causeway of land is manifestly ludicrous. At best it can only be regarded as a garbled version of the Prophet's actual statement. He may have poetically described the two spheres as the "wings" of the earth which, when they return, will be folded much as a bird's wings are folded against its body (see D&C 88:45). Dibble may have assumed that the two spheres—Enoch's city and the Lost Tribes?—were, therefore, literally connected to the earth in wing-like fashion. Oliver B. Huntington said Joseph Smith made a similar statement to that of Dibble's to him; see The Young Woman's Journal, April, 1894, 264.
63 The inhabitants, if any, of the south sphere were not identified. It was speculated that Enoch's city is located there (see *Times and Seasons,* 4:306-07).
64 A similar diagram was produced by Oliver B. Huntington which shows the two objects to be more like bulges or elongations of the earth than separate spheres (see The Young Woman's Journal, 3 [Mar. 1892], 264).
65 Letter of W.W. Phelps to Oliver Cowdery, *Messenger and Advocate,* vol. II, no. 1, (October, 1835), 94. See Jacob 5:13, 14,19, 38, 39. Oliver B. Huntington wrote with hyperbole, "Men have lost millions of dollars, and hundreds of lives to find a country beyond the north pole; and they will yet find that country—a warm, fruitful country, inhabited by the ten tribes of Israel, a country divided by a river, on one side of which lives the half tribe of Manasseh, which is more numerous than all the others. So said the Prophet. At the same time he described the shape of the earth at the poles as being a rounded elongation, and drew a diagram in this form: . . .He quoted scripture in proof of his theory which says that 'the earth flieth upon its wings in the midst of the creations of God. . .'" (The Young Woman's Journal, March, 1892, 263-64).

66 See JD, 18:26;19:171,173.
67 Smith and Sjodahl, 844.
69 *Defense of the Faith and the Saints,* 2:479-80. This position was also advocated in extensive treatments of the subject by James H. Anderson, *God's Covenant Race,* and Earnest L. Whitehead, *The House of Israel.*
70 See *Mormon Doctrine,* 455-58.
71 *A New Witness for the Articles of Faith,* 429, 520-21, 542.
72 It would be unprecedented because a similar miracle is nowhere recorded in Scripture. It would be uncharacteristic because it would interfere with the prevailing natural order over an extended period of time and constitute a control over human activities on a scale inconsistent with the Lord's established ways.
73 In commenting on this passage, Joseph Smith wrote: "It has been conjectured. . .[that] the ten tribes have been led away into some unknown regions of the north. Let this be as it may, the prophecy I have just quoted 'will fetch them,' in the last days, and place them in the land which their fathers possessed" (TPJS, 85).
74 Joseph Smith defined "outmost parts of heaven" as being "the breadth of the earth." Both phrases are ambiguous.
75 On the other hand, there is not a single, explicit statement by, or attributed to, Joseph Smith to the effect that the Lost Tribes are scattered among the nations. All of the Prophet's remarks—direct and hearsay—sustain the view that they are an integral group lost to the world.
76 Journal of Charles L. Walker, 2:505.
77 Minutes of the "School of the Prophets," (17 August 1872) Parowan, Utah, 156-57; emphasis added; see *Doctrines of Salvation,* 3:253.
78 "Recollections of the Prophet Joseph Smith," *Juvenile Instructor,* 27 (1 June 1892), 344.
79 *Millennial Star,* I (February, 1841) 258. At the time of Jesus' crucifixion, "there were some who were carried away in the whirlwind; and whither they went no man knoweth, save they know that they were carried away" (3 Ne. 8:16).
80 Robert W. Smith, *The Last Days,* 77. The statement of Homer M. Brown was made in the presence of Theodore Tobiason and Israel Call. Elder Brown was serving as a patriarch in the Granite Stake of the LDS Church in Salt Lake City, Utah at the time.
81 The statement that the earth is now ninety times smaller than its original size may have been a misunderstanding on Walker's part since another account indicates Eliza R. Snow said it was nine times smaller. Parley P. Pratt declared it to be several times smaller.
82 Church Archives, spelling and punctuation corrected.

Chapter Nine

Earth in The Meridian of Time

Four thousand years after the Fall, the Creator of heaven and Earth walked upon his footstool for the first and only time as a mortal being. He remained thirty-three years. His coming marked the beginning of the gospel dispensation of the meridian of time[1] in which three representative bodies of the house of Israel—the Jews in Palestine, the mixed remnant of Joseph and Judah in America, and the Lost Tribes in their unknown location—would be visited in turn by the Shepherd of Israel in an effort to gather them into one spiritual fold.

Signs of Jesus' Birth

Universal Light

The birth of Jesus Christ was accompanied by a most remarkable phenomenon in the Western Hemisphere: a period of approximately thirty-six hours of continuous daylight. Five years earlier, Samuel the Lamanite had prophesied:

> And behold, this will I give unto you for a sign at the time of his coming; for behold, there shall be great lights in heaven, insomuch that in the night before he cometh there shall be no darkness, insomuch that it shall appear unto man as if it were day. Therefore, there shall be one day and a night and a day, as if it were one day and there were no night; and

this shall be unto you for a sign; for ye shall know of the rising of the sun and also of its setting; therefore they shall know of a surety that there shall be two days and a night; nevertheless the night shall not be darkened; and it shall be the night before he is born (Hel. 14:3-4).

Since the sun followed its normal course, rising and setting throughout the globe, *the entire planet was bathed in light for a number of hours.* This unprecedented miracle signified the birth of him who, thirty-three years later, testified to the Jews of its spiritual meaning: "I am the light of the world: he that followeth me shall not walk in darkness, but shall have the light of life" (John 8:12).

A New Star

Samuel further prophesied: "And behold, there shall be a new star arise, *such an one as ye never have beheld;* and this also shall be a sign unto you. And behold this is not all, there shall be many signs and wonders in heaven" (Hel. 14:5-6).[2] Observed in the heavens of both the Eastern and Western Hemispheres, the new star was a fitting herald of the "bright and Morning Star" who will usher in the perfect day of millennial peace.

Of the four gospel writers, only Matthew mentions this star, telling us that "wise men" (learned sages from Persia) had seen it "in the east" and recognized it as "*his* star," meaning the star of the promised Messiah. Journeying to Jerusalem, they were interrogated by Herod as to exactly when the star first appeared. Herod, informed of Micah's prophecy,[3] then sent the Persians to Bethlehem—a distance of six miles. And "the star which they had seen in the East, *went before them,* till it came and stood over where the young *child* was.[4] When they saw the star they rejoiced with exceeding great joy" (Matt. 2:1-12). The account is puzzling.[5] How did these Persians identify this star with the

Messiah when the Jews did not? How could a star move and pinpoint a single house? Was it a true star, a sun, or was it something else? Samuel tells us that it was unlike any star ever seen before.[6] What set it apart from all others? Was it new-born, created for the occasion? Or was it an ancient visitor from other heavens? Since it was "his star"—seen in the east—announcing the Messiah's birth, it may have been the sign of the Son of Man—which will also be seen in the east—announcing the Messiah's return! Joseph Smith said this sign would be mistaken for a "planet, a comet, etc.,"[7] indicating it would appear to be a natural phenomenon. Of course, this is conjecture, but perhaps the Christmas star will appear again.

SIGNS OF JESUS' DEATH

The Redeemer Suffers

The rejection and condemnation of the Messiah by a conspiratorial Jewish hierarchy doomed his efforts to gather the house of Israel unto himself. Jesus then confronted his next mission; one performed at a fearful price: the Atonement. He told his disciples: "But I have a baptism to be baptized with; and how am I straitened till it be accomplished!" (Luke 12:50). Although he had been baptized by John in the waters of Jordan, the Son of Man was to endure a second, infinitely painful immersion when he descended below all things in Gethsemane. Trembling under the combined weight of humanity's guilt in this eternity, he bled at every pore in the bloody baptism of his incomprehensible sacrifice.

> For behold, I God, have suffered these things for all, that they might not suffer if they would repent. But if they would not repent they must suffer even as I; which suffering caused myself, even God, the greatest of all, to tremble because of

pain, and to bleed at every pore, and to suffer both body and spirit and shrink.... (D&C 19:16-18; see Luke 22:44).

Thereafter, in an ironic act of gratuitous savagery, the thorns that marred the planet following Adam's transgression were fashioned into a mock crown by Jesus' Roman tormentors and viciously pressed down upon his brow.[8] As he hung upon the cross of Golgotha, he wore the cruel symbol of that fallen world he had come to redeem! His impalement on the "tree" provided the visible tokens of his spiritual crucifixion in Gethsemane—spirit and element, the soul of the Redeemer, had been offered upon the altar of sacrifice to meet the law's demands. He had made himself "a curse for us" (Gal. 3:13). The cry: "Eli, Eli, lama sabachthani"— "My God, my God, why hast thou forsaken me?"—was not so much for himself as it was for mankind. For without Jesus Christ all mankind would be truly lost and God forsaken.[9] The crucified Son of Man not only personified the moral pains of the whole human family, but his scarred body bore the marks of his redemptive mission in their behalf.

The Crucifixion—Earth Mourns

Jesus Christ, the Spirit of truth, is in all and through all things.[10] When he trembled under the agonies of Gethsemane and Golgotha, all nature responded with empathic pain. This statement is not intended as poetic hyperbole. So perfectly attuned is the Creator to his creations that when certain Pharisees objected to Jesus' disciples shouting, "Blessed be the king that cometh in the name of the Lord," Jesus responded: "I tell you that, if these should hold their peace, the stones would immediately cry out" (Luke 19:40). If failure to acknowledge the kingship of Christ would have provoked the very stones to break their silence, would nature have remained indifferent to his death?"[11]

More than three thousand years earlier, Enoch, beholding the crucifixion in vision, heard a voice from the depths of Earth cry out:

> Wo, wo is me, the mother of men; I am pained, I am weary, because of the wickedness of my children. When shall I rest, and be cleansed from the filthiness which is gone forth out of me? When will my Creator sanctify me, that I may rest and righteousness for a season abide upon my face? (Moses 7:48).[12]

The Creator's crucifixion was the supreme offense against the earth whose elements had provided his very flesh and blood. Is it any wonder she reacted to this outrage as to no other? This ancient Patriarch comprehended and empathized with the emotions of Mother Earth: "And when Enoch *heard* the earth mourn, he wept, and cried unto the Lord, saying: O Lord, wilt thou not have compassion upon the earth?" (Moses 7:49). Enoch "beheld the Son of Man lifted up on the cross...And he heard a loud voice; and the heavens were veiled; and all the creations of God mourned; and the earth groaned" (Moses 7:55-56). Zenos, a prophet in ancient Israel, confirmed Enoch's vision when he wrote, "the rocks of the earth must rend; and because of the groanings of the earth, many of the kings of the isles of the sea shall be wrought upon by the Spirit of God, to exclaim: *The God of nature suffers*" (1 Ne. 19:12; see Rom. 8:19-23).[13]

Darkness Covers the Earth

As light accompanied Jesus' birth, so darkness accompanied his death. This too, was foretold by Samuel, the Lamanite prophet:

> But behold...in that day that he shall suffer death the sun shall be darkened and refuse to give his light unto you; and also the moon and the stars; and there shall be no light upon the face of this land, even from the time that he shall suffer death, for the space of three days, to the time that he shall rise again from the dead (Hel. 14:20).

Thus, the house of Israel in America again beheld a unique sign. For a three day period it experienced total, overwhelming darkness. In Jerusalem. during the period of the crucifixion, there was also a brief, three hour, period of darkness from 12 noon until 3 P.M.

> And it was about the sixth hour, and there was a darkness over all the earth until the ninth hour. And the sun was darkened. and the veil of the temple was rent in the midst (Luke 23:44-45).[14]

The Veil of the Temple

When the great veil of the temple in Jerusalem was torn in two from top to bottom, it exposed the Holy of Holies, the most sacred site in Jewry. The significance of this divine act, for it was a divine act, symbolized the end of the Mosaic covenant—the Old Testament—and the beginning of the new and everlasting covenant of the gospel—the New Testament. Even as Moses' smashing of the original tablets containing the higher law ended ancient Israel's abortive gospel dispensation, so did the rending of the temple veil signify the end of Israel's subjection to the lesser law of carnal commandments.[15] It also marked the end, as prophesied by Jacob, of Judah's spiritual primacy over Israel.[16] Most importantly, it meant that Jesus Christ had, by the sacrifice of his own divine life, entered the presence of the Majesty on High bearing the blood of the Atonement. This he did once and for all—thereby ending forever the need for, or the legitimacy of, the annual ritual of the levitical high priest on the day of atonement.[17] Lastly, it meant that no longer was a sinful, human, high priest needed to intercede for other sinners. All could enter the "holy of holies" in prayer and supplication bearing, not the blood of an animal in futile quest of forgiveness, but the sanctifying blood of "God himself." We approach the throne of grace—not in our own unworthy name—but in the name of Jesus Christ, the Father's sinless and eternal High Priest.

Judgment on American Israel

> For of him unto whom much is given much is required; and he who sins against the greater light shall receive the greater condemnation (D&C 82:3).

This principle applied to the house of Israel in America.[18] Of all mankind on the earth at the time, they possessed the greatest light. They had been led by prophets for six centuries. They had known of Christ and made solemn covenants with him. His Church, beginning with Alma, had been established among them for more than a century. They had been given a heaven-inspired constitutional government which assured them of both freedom and justice.[19] But in time those who chose evil became more numerous than those who chose good.[20] Apostasy became rampant; the government hopelessly corrupt.[21] They were "ripe for destruction."

And it came with a vengeance! The signs of the death of Christ were both graphic and consuming, taking the form of unparalleled natural disasters in the Western Hemisphere, where dwelt the Lehite-Mulekite nations.[22] However, these calamities did not come unannounced; nearly fifty years before they occurred, Samuel the Lamanite prophet foretold them in detail.[23] He prophesied that at the time Christ "yielded up the ghost," for three days—from his death to his resurrection—the light of the sun, moon, and stars would be blotted out. He further declared there would be "thunderings and lightnings for the space of many hours, and the earth shall shake and tremble," and the solid rocks *"above the earth and beneath. . .shall be broken up."* Also, there were to be "great tempests" and "many mountains laid low, like unto a valley" and many valleys "which *shall become mountains,* whose height is great." Finally, highways were to be "broken up," and many cities destroyed (Hel. 14:20-24).[24]

The prophet Mormon provides a graphic confirmation of these predicted judgments:

> . . .there arose a great storm, such an one as never had been known in all the land. And there was also a great and terrible tempest; and there was terrible thunder, insomuch that it did shake the whole earth as if it was about to divide asunder. And there were exceedingly sharp lightnings, such as never had been known in all the land. (3 Ne. 8:5-7).

Cities were hurled into the angry maw of the sea or inundated as its waters poured in upon them. Others disappeared when the earth parted to swallow them up. Still others became fiery ruins or were leveled to dust-laden rubble by the violence of rock crashing on rock. A great mountain formed a natural tombstone over the city of Moronihah when an incredible mass of earth was carried up and over it in one gigantic surge. Northward, the "whole face of the land was changed," making it impossible to determine its original geologic features. All of this in a matter of hours! Then came the darkness—darkness so thick, so impenetrable as to defy every effort to produce light.[25] The darkness which blanketed those ancient lands for three days was both a symbol of the spiritual darkness of those who perished and a reminder to us all of what our fate would be were it not for Jesus Christ, the Light of the world. When Christ died for sinners, many sinners died with him.

However, the calamities which befell the Lehite-Mulekite nation were not merely punitive in nature: they also served to purge the land of those who were hardened in sin and rebellion against God, having stoned the prophets sent to save them. Thus, only the more righteous survived the holocaust.[26]

The cleansing of that society paved the way for the personal visitation of the resurrected Christ some months later. Further, by removing the incorrigibles, it became possible for the Lord to establish his Church and kingdom among the Nephites in greater fullness than ever before. A new foundation was laid in righteousness upon which a Zion-like order was built which thrived for more than a century and a half. The spirit of the Priesthood dictated all that Christ did: in using the forces of nature against them he reproved American-Israel with sharpness; but in blessing them with his presence and his kingdom, he showed forth an increase of love.[27]

The judgments which were meted out on the Lehite-Mulekite nation were prophetic of the judgments which will be meted out on the entire world in these last days. Indeed, the accounts in the books of Helaman and Third Nephi constitute a prophetic blueprint of things to come in America and the world during the period leading up to the return of the Son of Man. The Book of Mormon is not only a witness for Christ, it is a voice of warning to all mankind, a voice which will be largely and tragically ignored.

The Scandal of Eternity

In crucifying their Messiah, Jesus' enemies among the Jews did what no other men would have done; for as the prophet Jacob noted, "there is none other nation on earth that would crucify their God" (2 Ne. 10:3).[28] Jesus came to Jerusalem—the one place on Earth where such a thing could happen. But more than this, he came to the one world in this eternity where such a thing could happen. No other earth harbored such appalling wickedness, such incredible insensitivity to the Spirit of God as did this beleaguered planet. Sin plummeted to its nadir in Jerusalem two thousand years ago. Glorious and beautiful in the beginning, Earth had given birth to the best and the worst of men—Christ and his prophets, Cain and the crucifiers of God. The Judge of all handed down the ultimate indictment when he told Enoch:

> Wherefore, I can stretch forth mine hands and hold all the creations which I have made; and mine eye can pierce them also, and among all the workmanship of mine hands there has not been so great wickedness as among thy brethren (Moses 7:36).

"Perhaps this is the reason Jesus Christ was sent here instead of some other world," wrote Joseph Fielding Smith, "for in some other world they would not have crucified Him, and His presence was needed here because of the extreme wickedness of the inhabitants of this earth."[29] What an ignominious distinction! Man has turned this planet into the scandal of eternity. He has made it a moral slum, unworthy of even the feet of an angel in his glory.[30]

Yet, it is a truism that neither men nor worlds can rise higher than they are capable of descending. The power to do good is proportionate to the power to do evil. Thus, Earth, having been capacitated to fall to the depths of degradation, will rise to the heights of glory. For as Brigham Young supposed:

> God never organized an earth and peopled it that was ever reduced to a lower state of darkness, sin and ignorance than this. I suppose this is one of the lowest kingdoms that ever the Lord Almighty created, and on that account is capable of becoming exalted to be one of the highest kingdoms that has ever had an exaltation in all the eternities (JD, 10:175).

Creation Travails

Earth, like a grieving widow, mourned the death of her Creator. She is in the dark hours before her deliverance. In prophesying of these latter-days, Isaiah wrote: "The earth mourneth and fadeth away, the world languisheth and fadeth away, the haughty people of the earth do languish" (Isa. 24:4). Neither Earth nor the Lord's people can be truly happy with the status quo.[31] Both man and Earth were created in glory for glory. Both fell to achieve a greater glory. For this reason, Earth and the sons and daughters of God who dwell upon her are in a state of anxious anticipation.[32] The apostle Paul felt that all creation was poised on the hope of Christ's emancipating return: "For we know that the whole creation groaneth and travaileth in pain together until now" (Rom. 8:22). Creation's pain, suffered in consequence of the Fall and the Atonement, reached its climax in the meridian of time—the spiritual nadir of human history.

Earth remembers her Lord. She not only bears record of him as Creator, but also testifies of him as Savior and Redeemer. His sacrifice is writ large upon the face of this planet in the language of nature. It is memorialized by scarred landscapes. It is witnessed by the agonized formations of fragmented rock and twisted mountains. It is spoken of by divided lands and lands beneath the sea. The testimony is there for every seeing eye and every discerning heart. It was meant to be so; the most momentous event in Earth's history was not to be forgotten or left without a memorial.

Notes

1 The meridian of time is the dividing point of time as we know it—the watershed of the temporal existence of Earth and man. In it the greatest events since the Fall occurred—the atonement and resurrection of Jesus Christ (see D&C 20:25-26).
2 See 3 Ne. 1:21 ; Matt. 1:10; Rev. 22:16. Samuel's prophecies are unique to the Book of Mormon. The heavenly signs were specifically prepared for the scattered house of Israel, not for the Gentiles. Appropriate signs were undoubtedly seen on other worlds as well.
3 See Micah 5:2.
4 According to Luke's account, Jesus was in Nazareth within three months of his birth. Matthew's account describes Jesus as a young child living in a "house" in Bethlehem when the wise men arrived. Upon their departure, Jesus was taken to Egypt. The apparent contradictions between the two accounts have never been satisfactorily resolved.
5 Most Biblical scholars regard the miraculous elements in the birth stories of Jesus in Matthew and Luke as myths invented to enhance claims to his divinity.
6 Astronomers typically explain the "Christmas star" as a fortuitous conjunction of several planets. However, the Book of Mormon indicates it was a solitary star.
7 TPJS, 287.
8 See JD, 4:199.
9 Ps. 22:1; Matt. 27:46; 2 Ne. 9:75-9.
10 See D&C 63:59; 84:45-46; 93:2, 9
11 See 3 Ne. 8:19.
12 Orson Pratt said: "When that Spirit, which is thus diffused through all the materials of nature, undertakes to converse with the minds of men, it converses in a different kind of language from that we use in our imperfect state. It communicates ideas more rapidly—more fully, and unfolds a world of knowledge in a moment" (JD, 3:102-03).
13 See 3 Ne. 8:19.
14 See Matt. 27:51; Mark 15:33, 38.
15 See Matt. 27:51; Mark 15:33; Luke 23:44; JST Ex. 34:1-2.
16 See Gen. 49:10.
17 See Heb. 9:11-28.
18 God's labor was with Israel, not the Gentile nations whose opportunity to receive the Gospel was yet in the future. The calamities associated with Christ's death were, therefore, directed against the rebellious of Israel, not the world at large (see 1 Ne. 19:11).

19 See Mosiah 29:27; Alma 10:19; Hel. 5:2.
20 See Hel. 4:22-23; 7:4-5; 3 Ne. 6:27-30.
21 Upheavals were experienced even on the isles of the sea; see 1 Ne. 19:11-12.
22 More than five hundred years earlier, Nephi beheld these destructions in vision (see 1 Ne.12:4-5).
23 The chaos which accompanied the death of the Creator reminds us of the chaotic state all nature would be in were it not for him.
24 See3 Ne. 8:21-22.
25 See 3 Ne. 10:11-13.
26 See D&C 121:43.
27 This indictment applies to the religious leaders and hierarchy who actually plotted his death, not to the Jewish people as a whole. Nevertheless, as Jesus said, "if the blind lead the blind, both shall fall into the ditch" (Matt. 15:14).
28 *The Signs of the Times,* 12.
29 This appears to be literally true since no authentic account of a glorious angelic visitation describes the messenger as touching the ground; see D&C 110:2; JS-H, 2:17, 30; Rev.1:15.
30 Brigham Young maintained, "happiness that is real and lasting in its nature cannot be enjoyed by mortals, for it is altogether out of keeping with this transitory state" (JD, 1:2; see D&C 101:36; Alma 37:45).
31 See John 17:14-17; Heb. 11:13-16; D&C 45:11-14.

Chapter Ten

Earth in Latter Days

Time is running out for this world. We are living in the closing moments of Earth's sixth temporal day—the day before the day on which Jesus Christ will return in judgment and the world as we know it will be no more.

Q. What are we to understand by the book which John saw, which sealed on the back with seven seals?
A. We are to understand that it contains the revealed will, mysteries and works of God; the hidden things of his economy concerning this earth during the seven thousand years of its continuance, or its temporal existence (D&C 77:6).

Hence, Joseph Smith's statement: "The world has had a fair trial for six thousand years; the Lord will try the seventh thousand Himself" (TPJS, 252). We are living in that period of human history called "the Saturday evening of time."[1] Soon there will be "time no longer" insofar as this fallen world and Satan (its alien god) are concerned.[2]

> Behold, now it is called today until the coming of the Son of Man. . . . .For after today cometh the burning—this is speaking after the manner of the Lord—for verily I say, tomorrow all the proud and they that do wickedly shall be as stubble; and I will burn them up, for I am the Lord of Hosts; and I will not spare any that remain in Babylon (D&C 64:23-24; see 45:6).[3]

But must the world end? Are there not enough righteous souls among us to save it? The last days are not a Sodom and Gomorrah situation; for several reasons, ten times ten million righteous souls in "Sodom" cannot prevent the end of the world from taking place. Ours is a fallen and, therefore, a temporary world; it was never meant to endure. It must pass away so that new heavens and a new earth endowed with terrestrial glory may take its place. Only then will Satan be bound for a thousand years. Only then will Earth have its Sabbath day. Only then will the Son of Man return and establish the fullness of the kingdom of God. Only then will the final preparations for the immortality and eternal life of mankind be completed. This world must come to an end so that the work and the glory of God may come to pass.

Withdrawal of the Spirit

An unperceived sign of the approaching end of this telestial order is the withdrawal of the Spirit of the Lord, the fountainhead of those natural forces man has identified, the divine mortar which binds all creation together into one orderly, harmonious structure. To the degree the Spirit is absent, chaos is present. The devils in hell, devoid of the Spirit, are not sane, for insanity is an expression of chaos. Where the Spirit is not, there will be a certain madness—the insanity of meaningless, uncomprehending existence. Simply stated, the less the light, the greater the darkness.

The Almighty's patience is not limitless; when men knowingly and willfully reject him to the spiritual point of no return—the point where their cup of iniquity is hopelessly full—he has no choice but to reject them. The Prophet warned, "God passes over no man's sins, but visits them with correction, and if His children will not repent of their sins He will discard them" (TPJS, 189).[4] The sun that blesses also burns; the rains that bring life can sweep it away. Both reveal God; both are valid expressions of his holy nature; both exemplify the two roads which lie before all men. In a letter to Nancy Rigdon, the Prophet wrote:

> Our heavenly Father is more liberal in His views, and boundless in His mercies and blessings, than we are ready to believe or receive; and, at the same time, is more terrible to the workers of iniquity, more awful in the executions of His punishments, and more ready to detect every false way, than we are apt to suppose Him to be (TPJS, 257).[5]

The prophesied judgments of the last days constitute the Almighty's last reminder to the world of this fact.

"For the Spirit of the Lord will not always strive with man. And when the Spirit ceaseth to strive with man then cometh speedy destruction, and this grieveth my soul" (2 Ne. 26:11). Proof of this statement by Nephi is found in Moroni's account of the Jaredites: "The Spirit of the Lord had ceased striving with them, and Satan had full power over the hearts of the people; for they were given up unto the hardness of their hearts, and the blindness of their minds" (Ether 15:19). In August, 1831, Jesus Christ said: "I, the Lord, am angry with the wicked; I am holding my Spirit from the inhabitants of the earth" (D&C 63:32). In January 1833 the Prophet alluded to this revelation and said that the withdrawal of the Spirit was an obvious fact:

> . . .the governments of the earth are thrown into confusion and division; and *Destruction,* to the eye of the spiritual beholder, seems to be written by the finger of an invisible hand, in large capitals, upon almost every thing we behold (TPJS, 16; emphasis in original).[6]

This generation is witnessing the acceleration of this withdrawal process. The "great apostasy" is greater now than it has ever been. To the distortion of Jesus' doctrines has now been added the rejection of his moral teachings. The irony of our times is that mankind's astonishing scientific and technological progress is inverse to our equally astonishing moral and spiritual decline. As our mastery of the physical order increases, our commitment to the moral order diminishes. We enjoy the illusion of progress at the very time we are descending into spiritual barbarism.

This retreat from both public and private morality has become a cliche of the times. The Media—newspapers, books, magazines, motion pictures, television, and now, the internet—spew forth the new hedonism. They have been joined by public schools, universities, and agencies of government in initiating programs and policies directly inimical to personal virtue and moral integrity. We have grown accustomed to the face of deceit, dishonesty, infidelity, immorality, gross perversions, vile language, the slaughter of the unborn, outrageous political knavery, etc. Jeremiah's description of his generation is applicable to our own: "Were they ashamed when they had committed abomination? nay they were not at all ashamed, neither could they blush" (Jer. 6:15; see 8:12).

The loss of the Spirit is the death of virtue. Few will mourn its passing as the last days draw to a close. Both Jew and Gentile "will be drunken with iniquity and all manner of abominations" (2 Ne. 27:1). Paul described the present generation:

> Men shall be lovers of their own selves, covetous, boasters, proud, blasphemers, disobedient to parents, unthankful, unholy, without natural affection, truce breakers, false accusers, incontinent, fierce, despisers of those that are good, traitors, heady, highminded, lovers of pleasures more than lovers of God; having a form of godliness, but denying the power thereof (2 Tim. 3:2-5).

In 1831, Jesus Christ summed up the modern age in one sentence: "Every man walketh in his own way, and after the image of his own god, whose image is in the likeness of the world" (D&C 1:16). Consider what has happened in the intervening one hundred sixty-nine years! How does the Lord view the world of 2000?

Man-Made Judgments

Man continues to be his own worst enemy. As the violation of moral law inevitably results in social upheaval, so does the violation of physical law produce natural upheavals. Consequently, there is a direct relationship between human behavior and the predicted judgments of the last days. In many instances, these judgments will be man-caused. A current example of this principle involves nuclear energy. Because of accidents and deliberate experimentation, agonizing sickness and death has befallen many thousands of helpless men, women, and children in areas of eastern Europe and western Asia now contaminated with lethal amounts of radioactive material.

Less dramatic is widespread deforestation and overgrazing which have led to the erosion of vital top soil, the formation of new desert land, and widespread flooding. The earth's waters have been similarly mistreated. The great fishing grounds have been decimated by over harvesting. Uncounted millions of tons of industrial and chemical pollutants have been callously discharged into the earth's rivers and oceans for many years. Concerned biologists are alarmed at the logarithmic speed with which the marine life cycle is being disrupted with a resultant loss of from thirty to fifty percent of both plant and fish life in the last few decades. These authorities warn that the waters, which cover seventy percent of the globe's surface, are relatively shallow when compared to the total bulk of the planet. Being so, they cannot be considered a "bottomless sewer." Indeed, the sea's ability to cleanse itself has long since been overwhelmed by staggering amounts of man-originating pollutants. The oceans are threatened with death.

The planet's stores of petroleum are also being consumed at the rate of billions of barrels each year. More importantly, the planet's aquifers are being steadily drained to make arid land productive, to provide for steadily growing and proliferating communities, and to meet the demands of industry. The result has been a steady and, in some places, an alarming drop in water tables. As with the rest of Earth's bounties, there is only a finite amount.

What will be the long range effects of these losses? Was Earth's inner wealth more closely associated with its dynamics than we have supposed? Did it serve to stabilize its internal forces? Is there a causal relationship between the draining of the planet's veins and arteries and the ever-increasing number of earthhquakes in "diverse places?"[7] Whatever the facts prove to be, one thing appears certain: our relentless consumption of these irreplaceable resources portends more than an inconvenience for future generations. It would be ironic if man's very effort to exploit this planet in the cause of modern civilization should become a direct factor in its end.

Of course, the Lord foresaw all of this and planned accordingly. It was foreknown that the world's last epoch would be the repository of all the genius, knowledge, power, and accomplishments of the race—the grand summation of the works of man, good and bad, for the past six thousand years. This epoch has its divine counterpart in the dispensation of the fullness of times in which all of the knowledge, keys, powers, principles, and ordinances of all previous dispensations are being gathered in one (see D&C 128:18). This is the climactic age of confrontation between the spiritual and materials orders, between the opposing kingdoms of God and Satan (see D&C 1:35-36; 38:12). Only one will survive.

The Lord's Sermons

The Wrath of God

The phrase, "wrath of God," engenders the image of a furiously offended Being. However, the term is simply an impressive way of personalizing the natural effects of broken law. Disobedience is the real precursor of the divine judgments which will be meted out by a just Father upon his rebellious children with ever-increasing frequency and severity. Because the Lord is obliged to honor the law he represents, his hands are tied; he cannot forever ignore the persistent defiance of the law and retain his own integrity. Divine justice could only be denied at the expense of God's position as God (see Alma 42:13, 22, 25). But the Almighty will stand guiltless before the bar of divine justice because he will have done everything possible to save mankind. Consider this remarkable passage:

> O, ye nations of the earth, how often would I have gathered you together as a hen gathereth her chickens under her wings, but ye would not! How oft have I calledupon you by the mouth of my servants, and by the ministering of angels, and by mine own voice, and by the voice of thunderings, and by the voice of lightnings, and by the voice of tempests, and by the voice of earthquakes, and great hailstorms, and by the voice of famines and pestilences of every kind, and by the great sound of a trump, and by the voice of judgment, and by the voice of mercy all the day long, and by the voice of glory and honor and the riches of eternal life, and would have saved you with an everlasting salvation, but ye would not! (D&C 43:24-25).

With these words Jesus Christ sums up his labors in behalf of mankind from the beginning to the end of time. Every possible approach—positive and negative—has been used within the limits of justice and mercy by God, angels, and men to restore the family of Man to his presence. "But ye would not!"

Now is the day of mercy when the Lord's servants are carrying his invitation to the wedding feast of salvation to the nations (see Matt. 22:1-14). However, comparatively few are responding to their message. Soon the voice of mercy will fall silent and the voice of justice will be heard as the testimonies of the servants are replaced by that of their Master. Heaven and Earth will combine in common witness as sign upon sign trumpets the fall of the kingdom of the devil and the end of his reign of blood and horror. As the year 1832 came to a close, the Lord revealed:

> For not many days hence and the earth shall tremble and reel to and fro as a drunken man; and the sun shall hide his face, and shall refuse to give light; and the moon shall be bathed in blood; and the stars shall become exceedingly angry, and shall cast themselves down as a fig that falleth from off a fig-tree. And after your testimony cometh wrath and indignation upon the people. For after your tesimony cometh the testimony of earthquakes. . . . And also cometh the testimony of the voice of thunderings, and the voice of lightnings, and the voice of tempests, and the voice of the waves of the sea heaving themselves beyond their bounds. And all things shall be in commotion (D&C 88:87-91).

Christ will speak to the nations through nature in upheaval. He will deliver his own sermons, not in words, but in awesome deeds whereby he will raise his voice in a last call to repentance—a call that will send the heavens and the earth into unprecedented commotion. Although natural disasters are strewn throughout history, their climactic forms have been reserved for these last days; the worst is yet to come!

Tempests in the form of tornadoes and hurricanes (typhoons) will be unleashed with mounting frequency and fury. Shattering electrical storms will rock the heavens with ear-tingling thunder and vivid lightning.[8] These will trigger widespread fires and devastating hailstorms.[9] One in particular—the last?—will apparently be world-wide in scope: "And there shall be a great hailstorm sent forth to destroy the crops of the earth" (D&C 29:16).

All these calamities, together with engulfing tidal waves, land-smothering volcanic eruptions, and deadly earthquakes, will combine their voices in crying, "Repent! Repent!"[10] In doing so, they will separate the "sheep from the goats," ridding the earth of many of those who persist in defiling it with their pollutions. Yet so universal will be the plagues of the last days that Joseph Smith warned: "it is a false idea that the Saints will escape all the judgments. . . many of the righteous shall fall a prey to disease, to pestilence, etc., by reason of the weakness of the flesh, and yet be saved in the kingdom of God."[11]

Zion and the Lord's Scourge

The Prophet Joseph prophesied, "God will gather out his saints from the Gentiles, and then comes desolation and destruction, and none can escape except the pure in heart who are gathered" (TPJS, 71). The redemption of Zion is the key to the major judgments of the latter-days which will begin in America and spread across the world. These judgments have in large measure been delayed because the building of the New Jerusalem has been delayed.[12] In 1831 the Saints were admonished:

> Go ye out of Babylon; gather ye out from among the nations. . . .Let them, therefore, who are among the Gentiles flee unto Zion (D&C 133:7, 12; see vss. 4-9).

Such was the grand theme in that early period.

Initially, Joseph Smith anticipated the successful gathering of the saints to the land of Zion, but for a number of reasons not germaine to this book, Zion was not established and the gathering to Missouri was eventually (1838) aborted.[13] Suffice to say, the saints were not sufficiently numerous, united, nor spiritually prepared for so grand an achievement.[14] One hundred sixty-six years have passed since "Zion's Camp," and Zion remains unredeemed. Nevertheless, while it was thought that the "center place"—New Jerusalem—would be established first, followed by stakes of Zion, the reverse occurred: stakes of Zion dot the world while the City remains a dream.

But it will arise in its time; all is on schedule. The relevant revelations are still valid, the Lord's conditions for success unchanged, and the necessity for the New Jerusalem as imperative as ever. It will emerge out of terrible turmoil "a land of peace, a city of refuge, a place of safety for the saints of the Most High God" (D&C 45:66). Together with its stakes, "holy places," it will be a beacon of light and hope to a world in darkness and despair— "the only people that shall not be at war one with another" (D&C 45:69).

In addition to war and desolation, the Doctrine and Covenants speaks of a strange malady which will strike the generation in which the "times of the Gentiles" are fulfilled:[15]

> And there shall be men standing in that generation, that shall not pass until they shall see an overflowing scourge; for a desolating sickness shall cover the land. But my disciples shall stand in holy places, and shall not be moved; but among the wicked, men shall lift up their voices and curse God and die. And there shall be earthquakes also in diverse places, and many desolations; yet men will harden their hearts against me, and they will take up the sword, one against another, and they will kill one another.[16]

This "desolating sickness" may be biological in nature, or it may result from exposure to lethal amounts of nuclear radiation. The unparalleled holocausts of the twentieth century should have taught us that the "natural man" remains carnal, sensual, and devilish; that Satan rages with ever-increasing power. Sadly, there is cause to fear the world will yet experience nuclear war.[17]

A most remarkable and terrible vision of latter-day conditions in the United States reminiscent of those depicted in Zechariah and the Doctrine and Covenants is recorded in the journal of Elder Wilford Woodruff.[18] A peculiar "disease" raged throughout the Utah Territory and the United States bringing agonizing death to millions, especially men:

> . . . I saw the roads full of people principally women, with just what they Could Carry in bundles on their backs travelling to the mountains on foot. . . .It was remarkable to me that there were so few men among them. . . .The rails looked rusty, and the road abandoned. . . .I continued eastward through Omaha and Council Bluffs which were full of disease and women every whare.

The author then describes appalling scenes of violent civil strife, familial murders, and worse. Washington D.C. was "a desolation," as was Baltimore, of which he writes:

> The waters of the Chesapeake and of the City were so stagnant and such a stench arose from them on account of the putrefaction of Dead bodies that the very smell Caused Death and that [what] was singular again I saw no men except they were dead, lying in the streets, and vary few women, and they were Crazy mad, and in a dying Condition. . . .And it was horrible, beyond description to look at.

He beheld similar scenes in Philadelphia, then continues:

> I next found myself in Broad way New York and here it seemed the people had done their best to overcome the Disease. But in wandering down Broadway I saw the bodies of Beautiful women lying stone dead, and others in a dying condition on the side walk. I saw men Crawl out of the Cellars and rob the dead bodies of the valuables they had on

and before they Could return to their coverts in the cellars they themselves would roll over a time or two and die in agony. On some of the back streets I saw Mothers kill their own Children and Eat raw flesh and then in a few minuts die themselves. Whareever I went I saw the same scenes of Horror and Desolation rapine and Death.

Seeing New York City utterly destroyed by fire,[19] the author comments:

> I Cannot paint in words the Horror that seemed to Encompass me around. It was beyond description or thought of man to Conceive. I supposed this was the End but I was here given to understand, that the same horror was being enacted all over the Country, North South East and West, that few were left alive. Still there were some.

The scene changed: "I saw the whole States of Missouri & Illinois and part of Iowa were a Complete wilderness with no living human being in them." He then beheld "Twelve men" consecrate the ground and lay the cornerstones of the New Jerusalem while "myriads of Angels" sang:

> Now is esstablished the Kingdom of our God and his Christ and He shall reign forever and Ever, and the Kingdom shall never be Thrown down for the Saints have overcome.

To what extent is this vision an accurate description of things to come? Time will tell. While we may consider it a nightmare, an aberration impossible of fulfillment, it is in harmony with a number of predictions made by latter-day prophets.[20] Joseph Smith himself foretold such devastations in the time of Zion's redemption:

> The time is soon coming when no man will have any peace but in Zion, and her stakes. I saw men hunting the lives of their own sons, and brother murdering brother, women

killing their own daughters, and daughters seeking the lives of their mothers. I saw armies arrayed against armies. I saw blood, desolation, fires (TPJS, 161).

In reference to the persecutors of the saints in Missouri, an 1839 revelation states:

> And not many years hence, that they and their posterity shall be swept from under heaven, saith God, that not one of them shall be left to stand by the wall (D&C 121:15; see 105:15).

The predicted judgments awaiting this nation serve to explain why the Lord said the saints would return to Missouri "to build up the *waste places* of Zion."[21] Following the establishment of Zion, the judgments of God will be withdrawn from a redeemed America. Thereafter they will be directed against the remaining Gentile nations. A graphic passage in an 1830 revelation describes a scourge which will befall the enemies of the Jews laying siege to Jerusalem in time to come.

> Wherefore, I the Lord God will send forth flies upon the face of the earth, which shall take hold of the inhabitants thereof, and shall eat their flesh, and shall cause maggots to come in upon them; And their tongues shall be stayed that they shall not utter against me; and their flesh shall fall from off their bones, and their eyes from their sockets; And it shall come to pass that the beasts of the forest and the fowls of the air shall devour them up.[22]

In 1950, President George Albert Smith uncharacteristically warned:

> . . .it is only a question of time, unless they repent of their sins and turn to God, that war will come, and not only war, but pestilence and other destruction, until the human family will disappear from the earth. . . .It will not be long until

calamities will overtake the human family unless there is speedy repentance. It will not be long before those who are scattered over the face of the earth by millions will die like flies because of what will come.[23]

There has not been "speedy repentance," to the contrary, there has been a steady retreat from even the unacceptable standards of fifty years ago. The world's cup of iniquity is rapidly filling.[24]

At the same time we read: "For thus saith the Lord, I will cut short my work in righteousness, for the days come that I will send forth judgment unto victory" (D&C 52:11). Every redemption is preceeded by the judgments of God. Needless to say, there is no such thing as a painless Second Coming.

RETURN OF THE LOST TRIBES

The place of the Lost Tribes—constituting the Northern Kingdom of Israel—in the prophetic scheme was pinpointed by the angel Moroni when instructing young Joseph Smith in 1823. After prophesying of the coming forth of the Book of Mormon, of the rise of the Church, and of the persecution which would attend the latter-day work, Moroni stated:

> Then will persecution rage more and more, for the iniquities of men shall be revealed, and those who are not built upon the Rock will seek to overthrow this church; but it will increase the more opposed, and spread farther and farther, increasing in knowledge till they shall be sanctified and receive an inheritance where the glory of God shall rest upon them; and when this takes place, and all things are prepared, the ten tribes of Israel will be revealed in the north country, whither they have been for a long season; and when this is fulfilled will be brought to pass the saying of the prophet—

'And the Redeemer shall come to Zion, and unto them that turn from transgression in Jacob, saith the Lord.'[25]

Following the judgments on America and the redemption of Zion—including the building of the city New Jerusalem—the Lost Tribes will come down from the land of the north. In doing so, they will unite with "Ephraim and his fellows" who are already gathered into the fold of Jesus Christ—his Church. In other words, those of the Northern Kingdom of Israel (the Lost Ten Tribes) will be gathered unto Christ some years before those of the Southern Kingdom (Judah). In redeeming the house of Jacob, the first (Judah), shall be last, and the last (Israel) shall be first. Thus, before Jesus Christ appears to the Gentiles at his world coming, he will minister first to the house of Israel (Ephraim) in America and, thereafter, to Judah in Palestine.

Keys for Leading Lost Tribes

Israel's ingathering is a vital part of the restitution of all things wherein, as Joseph Smith wrote: "a whole and complete and perfect union, and welding together of dispensations, and keys, and powers, and glories should take place, and be revealed from the days of Adam even to the present time" (D&C 128:18; see Eph. 1:9-10). This union must include the redemption of righteous Israel, both physically and spiritually. The authority for Ephraim, the elder brother of Israel,[26] to accomplish this labor was given to Joseph Smith and Oliver Cowdery in the Kirtland Temple on 3 April 1836 by Moses himself:

> Moses appeared before us, and committed unto us the keys of the gathering of Israel from the four parts of the earth, *and* the leading of the ten tribes from the land of the north (D&C 110:11).[27]

Note that the gathering of Israel from the four parts of Earth is distinct from and, to a degree, preliminary to the leading back of the Ten Tribes from the land of the north. Clearly, the two activities, while related, are not to be equated as one.[28]

John the Revelator

Jehovah's purpose in scattering Israel will become increasingly apparent as the end draws near. Israel is the spiritual "salt" of the human family—the means by which God's covenant with Abraham will be fulfilled.[29] However, in seasoning the world with the blood of Israel, it was neither necessary nor desirable that all of this "salt" be poured into the spiritual food of the nations. The needed Gentile-Israelite mixture did not require a total scattering of the twelve tribes; a remnant was to be preserved. A second effort is needed to redeem Israel because the first, that of Jesus, was unsuccessful. Therefore, "the Lord shall set his hand again the second time" to gather his people in these latter days (see 2 Ne. 29:1).

The apostle John's desire to continue to live and bring souls to Christ until his return was granted. He was to do a greater work and "prophesy before nations, kindreds, tongues and people."[30] In time, he became a translated being, no longer subject to the limitations of mortality.[31] The "greater work" assigned to John included the responsibility for overseeing the "gathering in" and "sealing up" of the twelve tribes of Israel preparatory to Christ's final coming.[32] This mission was symbolized in John's vision by the "little book" which, when he ate it, was sweet in his mouth but bitter in his stomach.[33] That is, the mission was sweet in anticipation, but difficult in realization.

John has been laboring among the nations for more than nineteen centuries. In doing so he has also ranged far and wide in ministering to the scattered peoples of Israel. In the absence of the Shepherd, John has become the guardian of his flock.

According to John Whitmer's report of the fourth general conference of the Church held June 3-6, 1831:

> The Spirit of the Lord fell upon Joseph in an unusual manner, and he prophesied that John the Revelator was then among the Ten Tribes of Israel who had been led away by Shalmaneser, king of Assyria,[34] to prepare them for their return from their long dispersion, to again possess the land of their fathers (DHC, 1:176).

Daniel Allen testified:

> I heard Joseph the Prophet say that he had seen John the Revelator and had a long conversation with him, who told him that he John was their leader, Prophet, Priest, and King, and said that he was preparing that people to return and further said there is a mighty host of us.[35]

Orson Pratt stated, "John the Revelator will be there, teaching, instructing and preparing them. . .for to him were given the keys for the gathering of Israel" (JD, 18:25-26).[36]

John's assignment to "minister for those who shall be heirs of salvation who dwell on the earth," does not necessarily preclude a labor to regions beyond the earth. In addressing the general conference of October 1840, the Prophet Joseph Smith referred to translated beings:

> . . .their place of habitation is that of the terrestrial order, and a place prepared for such characters He held in reserve to be ministering angels unto many planets (TPJS, 170).

If the Lost Tribes were transported to another world, the mission given John seemingly would require him to minister to them there, and the powers possessed by him as a translated being would enable him to do so.

The Return

The exodus of Israel from Egypt involved the mass movement of hundreds of thousands, if not several million, men, women, and children, together with their livestock and other possessions.[37] It was an undertaking of enormous proportions. Yet great as it was, the return of the Lost Tribes will cause it to pale by comparison. Jeremiah prophesied:

> Therefore, behold the days come, saith the Lord, that it shall no more be said, The Lord liveth, that brought up the children of Israel out of the land of Egypt; But, the Lord liveth, that brought up the children of Israel from the land of the north, and from all the lands whither he had driven them; and I will bring them again into their land that I gave unto their fathers (Jer. 16:14-15).[38]

The work of the "fishers and the hunters" in carrying the message of salvation to the nations and gathering out the believing blood of Israel— "one of a city, and two of a family"—is not a mass movement comparable to the exodus of ancient Israel. Only the return of a mighty multitude can satisfy Jeremiah's prophecy.[39] And such will be the case. The most unique and detailed revelation on the return of the Lost Tribes was appropriately given to Joseph Smith who holds the keys for their return:

> And they who are in the north countries shall come in remembrance before the Lord; and their prophets shall hear his voice, and shall no longer stay themselves; and they shall smite the rocks, and the ice[40] shall flow down at their presence. And an highway shall be cast up in the midst of the great deep. Their enemies shall become a prey unto them, And in the barren deserts there shall come forth pools of living water; and the parched ground shall no longer be a thirsty land. And they shall bring forth their rich treasures unto the children of Ephraim, my servants. And the boundaries of the everlasting hills shall tremble at their presence. And there

shall they fall down and be crowned with glory, even in Zion, by the hands of the servants of the Lord, even the children of Ephraim. And they shall be filled with songs of everlasting joy. Behold, this is the blessing of the everlasting God upon the tribes of Israel, and the richer blessing upon the head of Ephraim and his fellows (D&C 133:26-34).[41]

This categorical revelation is explicit on several points. The Lost Tribes will suddenly appear in power and glory somewhere in the north, causing a mighty upheaval in nature.[42] Their destination is America, the land of Zion, for, said Joseph Smith, "unto it all the tribes of Israel will come" (TPJS, 17).[43] They will be led by "their prophets" who are preparing them to rejoin the rest of the family of Israel and to receive the fullness of the Lord's blessings. Orson Pratt remarked:

> Do not think that we are the only people who will have Prophets. God is determined to raise up Prophets among that people, but he will not bestow upon them all the fulness of the blessings of the Priesthood. The fulness will be reserved to be given to them after they come to Zion. But Prophets will be among them while in the north, and a portion of the Priesthood will be there; and John the Revelator will be there (JD, 18:25).[44]

Wilford Woodruff agreed:

> There are Prophets among them, and by and by they will come along, and they will smite the rocks, and the moutains of ice will flow down at their presence, and a highway will be cast up before them, and they will come to Zion, receive their endowments, and be crowned under the hands of the children of Ephraim, and there are persons before me in this assembly to-day who will assist to give them their endowments (JD, 4:231-32).[45]

In 1880, twenty-three years later he testified:

> . . .all that God has said with regard to the ten tribes of Israel, strange as it may appear, will come to pass. They will, as has been said concerning them, smite the rock and the mountains of ice will flow before them, and a great highway will be cast up, and their enemies will become a prey to them; and their records, and their choice treasures they will bring to Zion. These things are as true as God lives (JD, 21:301; emphasis added).[46]

In 1920, Elder Melvin J. Ballard:

> My witness and testimony is that . . . even now their prophets prophesy of their deliverance and are preparing that people. The mountains of ice shall flow down before them, and a highway shall be cast up, and they shall come forth with songs of everlasting joy, to be crowned at the hands of the children of Ephraim (CR, Oct. 1920, 82).

In 1959, Elder Spencer W. Kimball:

> They will return with their prophets, and their sacred records will be a third witness for Christ (CR, Oct. 1959, 61).[47]

When the divine command is given, their prophets "shall no longer stay themselves:" they will descend from the "north countries." And even as the mountains shall flow down and the seas boil at the presence of the Son of Man in his glory, so will the glacial snows melt and their icy waters flow down at the presence of the returning hosts of Israel. The very manner in which they make their appearance will release an energy force sufficiently great to melt some of the frozen reaches of the north![48] A land bridge—a highway to Zion[49]—will miraculously span a portion of the Arctic Ocean enabling the tribes to journey southward into North America—for has not the Lord "made the depths of the sea a way for the ransomed to pass over?" (Isa. 51:10).[50]

It is almost unanimously agreed the Lost Tribes do not exist as an organized body on this planet. Consequently, if such a body exists, it must have an extra-terrestrial location. And if its situation is similar to that of Enoch's city, its return would be, to say the least, astonishing. The approach of an extra-terrestrial body of sufficient mass could affect the earth's orbital stability causing it to move erratically. From the standpoint of contemporary science, such a thing would be catastrophic in the extreme. However, the Almighty is fully cognizant of the physics involved; its accomplishment would be a miracle to everyone but himself.

Wandle Mace heard the Prophet Joseph Smith illustrate how the return of the Lost Tribes and Enoch's city would affect the planet:

> Some of you brethren have been coming up the river, on a Steamboat, and while seated at the table, the Steamboat runs against a snag upsetting the table and scattering the dishes. So it will be when these portions of earth return. It will make the Earth reel to and fro like a drunken man.[51]

The Prophet then quoted from Isaiah: "The earth shall reel to and fro like a drunkard, and shall be removed like a cottage; and the transgression thereof shall be heavy upon it; and it shall fall, and not rise again" (Isa. 24:20).[52] In 1840 Samuel H. Rogers recorded a similar explanation by Joseph Smith of the phenomenon:

> We also read in the Scriptures that the earth shall reel to and fro like a drunken man. What shall cause this earth to reel to and fro like a drunken man? We read that the stars shall fall to the earth like a fig falling from a fig tree. When these stars return to the place where they were taken from, it will cause the earth to reel to and fro. Not that the planets will come squarely against each other, in such a case both planets would be broken to pieces. But in their rolling motion they will

come together where they were taken from which will cause the earth to reel to and fro.[53]

Recall that Bathsheba W. Smith and Eliza R. Snow also testified to having heard Joseph Smith teach that the return of the Lost Tribes would cause, or be in connection with, the planet reeling like a drunken man. We have noted the testimonies of a number of the Prophet's friends and associates as to his views on the present whereabouts of Enoch's city and the Lost Tribes. There is no doubt but that their reunion with this planet would trigger just such a phenomenon.[54] Whatever causes the earth to reel to and fro, it will be of an unprecedented nature. The physical energy then exerted will be far more powerful than all of the earthquakes, volcanic eruptions, and other ongoing geologic upheavals of the last days combined!

Enemies A Prey

The return of the Lost Tribes will be accompanied by earthquakes: "the boundaries of the everlasting hills shall tremble at their presence." Orson Pratt predicted:

> ... this great Rocky Mountain range, extending from the Arctic regions south to the central portions of America, will tremble beneath the power of God at the approach of that people (JD, 18:24; see Ps. 114).

They will not be allowed to enter North America without opposition; "enemies"[55] seeking to repel them as invaders or simply desiring to exploit them will confront them. But the host of Israel will be invincible; by the power of the Priesthood they will vanquish their "prey" and proceed on to the land of Zion. In that day the United States will not be as it is now with its sprawling cities teeming with millions of inhabitants. It appears, the wickedness of men and the judgments of God will have combined to markedly reduce the population. In an 1833 letter to N.E. Seaton, editor of a Rochester, New York, newspaper Joseph Smith wrote:

> And now I am prepared to say by the authority of Jesus Christ, that not many years shall pass away before the United States shall present such a scene of bloodshed as has not a parallel in the history of our nation; pestilence, hail, famine, and earthquake will sweep the wicked of this generation from off the face of the land, to open and prepare the way for the return of the Lost Tribes of Israel from the north country (TPJS,17).[56]

The Civil War did not fulfill this prophecy. In a time to come natural disasters and disease combined with international and civil conflict, will apparently sweep millions from off the land. Isaiah prophecied that Israel would "break forth on the right hand and on the left;" and that their "seed" would "inherit the Gentiles and make the desolate cities to be inhabited" (Isa. 54:3).[57] This suggests that cities in America, depleted of their inhabitants, will become the dwelling places of the Lost Tribes. America is ignoring the clear warnings of The Book of Mormon against rejecting Jesus Christ, the God of the land; it faces his judgments.[58]

Time of Return

As we have seen early leaders of the Church, including Joseph Smith, believed the return of the Lost Tribes to be relatively imminent. This, because their return was to precede by some years the coming of Christ to the world at large. Reflecting this belief in 1857, Wilford Woodruff predicted, "there are persons before me in this assembly to-day, who will assist to give them their endowments" (JD, 4:232). In 1875 he again said, "those long Lost Tribes will come forth in your day and mine, if we live a few years longer" (JD, 18:38). Orson Pratt concurred:

> There are persons in this congregation who will be in the midst of Zion, when the ten tribes come to Zion from the north countries, and will assist in bestowing the blessings promised by the Almighty upon the heads of the tribes of Israel (JD, 18:25).

Their return was also predicted in the twentieth century. Elder James E. Talmage testified: "I say unto you there are those now living—aye, some here present—who shall live to read the records of the Lost Tribes of Israel."[59] And, in 1967, William J. Critchlow, Jr. suggested, "Some of my grandchildren may witness the return of the lost ten tribes" (CR, Oct. 1967, 88). It seems almost paradoxical that the nearer we get to the fulfillment of prophecy, the more cautious some become in affirming it. Still, as stated, before, the time factor in prophecy is the unknown factor. Knowing *what* is not necessarily knowing *when*. Not even Jesus knew when he would return.[60]

Rich Treasures

When the Lost Tribes return they shall "bring forth their rich treasures unto the children of Ephraim, my servants" (D&C 133:30). These "rich treasures" will doubtlessly include essential scriptural, genealogical and historical records, including an account of the Redeemer's ministry among them.[61] An April, 1845 proclamation by the Council of the Twelve to the world stated:

> . . .the ten tribes of Israel should also be revealed in the north country, together with their oracles and records, preparatory to their return, and to their union with Judah, no more to be separated.[62]

The fact that the tribes will return with their own scriptural record is a telling argument for their isolated state since such a volume does not exist in the known world. Note too, they bring their records with them; they are not produced subsequent to their return.[63] Thus, as part of the restitution of all things, the word of the Lord to the Jews (the Bible), to Joseph (the Book of Mormon), and to the Lost Tribes "shall be gathered in one" (2 Ne. 29:14). The record of the Lost Tribes, said Elder Spencer W. Kimball, constitutes "a third witness for Christ" (CR, Oct. 1959, 61). Through these three scriptures the ministry of the Holy One to scattered Israel will become one united testimony for all "who remain" to receive him. Joseph, the Prophet, asked:

> What was the object of gathering the Jews, or the people of God in any age of the world? . . .The main object was to build unto the Lord a house whereby He could reveal unto His people the ordinances of his House and the glories of His kingdom, and teach the people the way of salvation (TPJS, 308).

Thus, the House of the Lord will be the primary objective of the Lost Tribes in coming to Zion.[64] Having obtained the saving grace of the gospel, they will seek a fullness of exaltation. They will obtain it through Ephraim—the "first born" in the house of Israel. Through him the dreams of ancient Joseph will find ultimate fulfillment. The sheaves and the stars will bow before him in receiving the "blessings of the everlasting God."[65] They shall "fall down and be crowned with glory, even in Zion, by the hands of the servants of the Lord, even the children of Ephraim" (D&C 133:32). The Lost Tribes of Israel will be lost no more.

THE RETURN OF ENOCH'S ZION

Awesome indeed will be the glorious return to Earth of Enoch and his "City of Holiness, even ZION." Their appearing will be a heavenly sign of the imminence of the coming of the Son of Man, the Messiah—first to the Jews, and then to the world.[66] The Lord told Enoch that in the last days "righteousness and truth will I cause to sweep the earth as with a flood to gather out mine elect from the four quarters of the earth" (Moses 7:62). This was to be accomplished through the labors of the Restored Church of Jesus Christ. Predominantly the blood of Joseph through Ephraim and Manasseh, it is now engaged in a world-wide missionary effort to that end.

Zion Above, Zion Beneath

> The Lord hath gathered all things in one.
> The Lord hath brought down Zion from above.
> The Lord hath brought up Zion from beneath
> (D&C, 84:100).

Enoch learned from the Son of God himself that the saints of latter-days would also build a "Holy City"—Zion, a New Jerusalem—to which the elect would be gathered in anticipation of the Savior's return. These two Zions will become as one. And just as God "dwelt in the midst of Zion" anciently, so will God in the person of Jesus Christ dwell with his united people before his world coming. The Son of Man described the meeting of these Zion peoples in very human terms:

> Then shalt thou and all thy city meet them there, and we will receive them into our bosom, and they shall see us; and we will fall upon their necks, and they shall fall upon our

necks, and we will kiss each other. . . .and for the space of a thousand years the earth shall rest (Moses 7:63-64).

The use of the pronouns "we," "our" and "us" in this passage suggests the profound oneness of Jesus Christ and the people of Enoch. They fled to his presence when they left the Earth; they will be with him when he comes to his latter-day saints. The Zion of Enoch and the Zion of Joseph will be united forever as the exalted ones comprising "the church of Enoch, and of the Firstborn."[67]

Zion Expands

We have noted that the Prophet Joseph Smith is credited with teaching that Enoch's city would return to its original site, supposedly in the region of the Gulf of Mexico. This will place it in proximity to the New Jerusalem, the center place of the Zion of the latter-days, in the land of Missouri. The New Jerusalem will be magnified through the establishment of many Zion-like communities, thereby expanding into "the regions round about."[68] In time, this process will produce a perfect union of the two cities—the first and the last—one vast, magnificent land of Zion. Noah was told that the rainbow would betoken the union of these terrestrial societies:

> And the bow shall be in the cloud; and I will look upon it, that I may remember the everlasting covenant, which I made unto thy father Enoch; that, when men should keep all my commandments, Zion should again come on the earth, the city of Enoch which I have caught up unto myself. And this is mine everlasting covenant, that when thy posterity shall embrace the truth, and look upward, then shall Zion look downward, and all the heavens shall shake with gladness, and the earth shall tremble with joy; And the general assembly of the church of the first-born shall come down out of heaven, and possess the earth, and shall have place until the end

come. And this is mine everlasting covenant, which I made with thy father Enoch (JST, Gen. 9:21- 23).

Noah's righteous posterity will look "upward" when the Kingdom of God comes forth from The Church of Jesus Christ of Latter-day Saints and the Lord's people dwell in righteousness with an eye single to the glory of God. In other words, the Zion of Enoch will cleave to the Zion of Joseph when the Zion of Joseph becomes in very deed, "the pure in heart" (D&C 97:21).

Sign of the Son of Man

Following the negative signs associated with the sun, moon, and stars, the final heavenly token preceding the Lord's second advent will be the positive sign of the Son of Man. It will appear in the eastern sky throughout the earth. Jesus taught his disciples: "after the tribulation of those days, and the powers of the heavens shall be shaken, then shall appear the sign of the Son of Man in heaven; and then shall all the tribes of the earth mourn" (JST Matt. 24:37).[69] The Prophet Joseph Smith expanded on this statement which applies to events after the redemption of Zion:

> There will be wars and rumors of war, signs in the heavens above and on the earth beneath, the sun turned into darkness and the moon to blood, earthquakes in diverse places, the seas heaving beyond their bounds; then will appear one grand sign of the Son of Man in heaven. But what will the world do? They will say it is a planet, a comet, etc. But the Son of Man will come as [like] the sign of the coming of the Son of Man, which will be as the light of the morning cometh out of the east (TPJS, 286-87).

Wandle Mace heard the Prophet teach that the principle of the first being last and the last being first would also apply to the return of Enoch's city and the Lost Tribes: since Enoch's city was the first body taken from this planet, it will be the last body to return. He also quoted Joseph Smith as teaching that this climactic sign was "the return of the City of Enoch to the earth."[70] The fact that the sign of the Son of Man and the return of Enoch's Zion are both placed in the same prophetic time frame—being associated with the imminent world appearance of the Lord—is supportive of the argument they are one and the same.

It would be appropriate for ancient Zion to be the sign of him whose people they are. He has been their God and King for some five thousand years; they have walked together. Theirs has been a millennial-like relationship; they bask in terrestrial glory and will be reunited with the earth when it partakes of a like glory. Since all creation will be awaiting the coming of the Lord, Enoch's city will return to Earth prior to that climactic event.

Years before this grand union occurs the words of Isaiah will be fulfilled: "And the Redeemer shall come to Zion, and unto them that turn from transgression in Jacob, saith the Lord" (Isa. 59:20). This he will do that "my covenant people may be gathered in one in that day when I shall come to my temple" (D&C 42:36).[71] As he appeared to the Lehites at the temple in Bountiful, so will he appear to the Latter-day Saints at the temple in New Jerusalem. As he dwelt with his people in Enoch's City of Holiness before Earth's watery baptism, so will he dwell with his saints in the New Jerusalem before Earth's fiery spiritual baptism.[72]

Signs on Earth and in Heaven

The Half Hour of Silence

The opening of the "seventh seal" will begin with "silence in heaven about the space of half an hour" (Rev. 8:1).[73] The social disorders and natural upheavals which will occur in connection with Zion's redemption in the time of the sixth seal will foreshadow the final judgments associated with the redemption of Jerusalem following that silence in the time of the seventh seal. For some twenty-one years, during which the Lord will complete the gathering of the "wheat" into his garner, the "tares" will have a last opportunity to reflect upon and, hopefully, repent of their rebelliousness.[74] However, mankind in general will continue to be indifferent to heaven's pleas, being absorbed in its own self-generated pleasures and problems. Jesus compared those times to the closing moments of Noah's world when "they were eating and drinking, marrying and giving in marriage, and knew not till the flood came and took them all away" (JST, Matt. 24:45). Thus, following that interlude of relative peace, heaven's silence will be broken by the successive sounding of the trumpets of the seven angels of Revelation associated with the final series of judgments culminating in Christ's world appearance.[75]

Prior to that "great day," the heavens and the earth will abound with awesome portents.[76] The most vivid of these will be associated with the house of Israel: the redemption of Zion, the return of the Lost Tribes, the return of Enoch's city, and the deliverance of a remnant of Judah at Jerusalem. The redemption of Israel which began in the 19th century apparently will be fully realized in the 21st century. The signs accompanying this final in-gathering will, therefore, span the same time period. As we approach the grand denouement of prophecy, these signs will increase in frequency and intensity until they reach a climax in the heavens.

The Vanished Rainbow

Since the days of Noah, the rainbow has been a sign of the Lord's mercy which has been showered upon mankind for six thousand years. Now approaches the day of justice. When justice replaces mercy, the token of the rainbow will be seen no more in this planet's skies. Its departure will portend further tribulations and signify the imminent coming of the Son of Man. The Prophet Joseph Smith explained the significance of the rainbow for these latter days:

> I have asked the Lord concerning His coming; and while asking the Lord, He gave a sign and said, 'In the days of Noah I set a bow in the heavens as a sign and token that in any year that the bow should be seen the Lord would not come; but there should be seed time and harvest during that year: but when ever you see the bow withdrawn, it shall be a token that there shall be famine, pestilence, and great distress among the nations, and that the coming of the Messiah is not far distant (TPJS, 340-41; see 305).[77]

Having rejected the divine invitation to eternal life (as signified by the rainbow), mankind will experience a new sign in the form of drought with its grim attendants—famine and pestilence. As it was in the days of Noah, so shall it be in that day—with one major exception: men will not cower before flood-laden skies, instead they will face cloudless heavens and a merciless sun beating down upon scorched and barren lands—portents of the burning to come.

Stars Shall Fall

So violent and extensive shall be the upheavals which shake the planet in future days that those conditions which prevailed in the Western Hemisphere during Christ's crucifixion will be repeated: "There was thick darkness upon the face of the land"

(3 Ne. 8:20). The lights of heaven will go out: the "stars of heaven and the constellations thereof shall not give their light: the sun shall be darkened in his going forth, and the moon shall not cause her light to shine" (Isa. 13:10). Joel confirms Isaiah's prophecy times. First in connection with the redemption of Zion in America and thereafter in connection with Christ's deliverance of the Jews at Jerusalem.[78] It was in the context of Judah's deliverance that Jesus reiterated the words of Isaiah and Joel:

> And immediately after the tribulation of those days, the sun shall be darkened, and the moon shall not give her light, and the stars shall fall from heaven, and the powers of heaven shall be shaken (JST Matt. 24:34).[79]

This prophecy also appears once in the book of Revelation and four times in the Doctrine and Covenants.[80] Most commentators assume that these "stars" will consist of meteoric showers. However, meteorites do not constitute a unique sign; meteoric showers or "falling stars" have been observed throughout the ages. Impressive as they are, they are still too prosaic to qualify as distinct signs of the approaching return of the Son of Man. Then, too, he has declared that "the stars shall be hurled from their places" (D&C 133:49). Meteoroids are, by definition, placeless wanderers in space; they have no fixed orbits from which they could be hurled.

The prophecy only pertains to a select number of "stars" or heavenly bodies. John wrote, "the stars of heaven fell unto the earth, even as a fig tree casteth here untimely figs, when she is shaken of a mighty wind" (Rev. 6:13).[81] Untimely figs are those which ripen in the winter out of "due time". As such, they are comparatively few in number. So too, comparatively few stars will fall in connection with Christ's return. This view is confirmed in a modern revelation which states "some shall fall" (D&C 34:9).

It should also be noted that the stars seen by John fell to the earth. This would seem to support the theory of meteoric showers since the smallest known star is many times larger than this planet and would engulf it like a whale swallowing plankton. Joseph Fielding asked the question: "How can the stars fall from heaven to earth, when they (as far as we know) are much larger than the earth?" Parley P. Pratt, editor of the Millennial Star, responded:

> We are nowhere given to understand that all the stars will fall or even many of them: but only 'as a fig tree casteth her UNTIMELY figs when she is shaken with a mighty wind.' The stars which will fall to the earth, are fragments, which have been broken off from the earth from time to time, in the mighty convulsions of nature. Some in the days of Enoch, some perhaps in the days of Peleg, some with the ten tribes, and some at the crucifixion of the Messiah. These all must be restored again at the 'times of restitution of ALL THINGS.' This will restore the ten tribes of Israel; and also bring again Zion, even Enoch's city (*MS*, I [Feb. 1841], 258; emphasis in original).[82]

Defining John's "stars" as returning portions of the planet does not preclude the participation of true stars in the signs of Christ's coming.[83] What is significant is that Parley P. Pratt, an associate of Joseph Smith, understood that Enoch's city and the Lost Tribes would constitute two of those "stars" John saw "fall to the earth."[84]

The Earth Shall Reel

Thus, signs in the heavens will be concurrent with unprecedented upheavals on the earth. The Lord revealed:

> . . .not many days hence and the earth shall tremble and reel to and fro as a drunken man; and the sun shall hide his

face, and shall refuse to give light; and the moon shall be bathed in blood; and the stars shall become exceedingly angry, and shall cast themselves down as a fig that falleth from off a fig-tree (D&C 88:87).[85]

Recall that Joseph Smith was quoted as associating the Earth's reeling with the Lost Tribes and with Enoch's city. However, in this instance, it pertains to the time of the redemption of the tribe of Judah. A companion revelation states:

> Then shall the arm of the Lord fall upon the nations. And then shall the Lord set his foot upon this mount, and it shall cleave in twain, and the earth shall tremble, and reel to and fro, and the heavens also shall shake. And the Lord shall utter his voice, and all the ends of the earth shall hear it; and the nations of the earth shall mourn, and they that have laughed shall see their folly (D&C 45:47-49).[86]

This particular reeling of the planet is also associated with the battle of Armageddon— "the battle of that great day of God Almighty" (Rev. 16:14). When the armies of the Gentile nations are about to overwhelm the Jews in Jerusalem after a siege of more than three years, the Messiah will descend to the Mount of Olives fulfilling Joel's prophecy: "Multitudes, multitudes in the valley of decision: for the day of the Lord is near in the valley of decision" (Joel 3:14). The divine destruction of the armies of "Gog" marshalled against the Jews will be accompanied by the mightiest earthquake in the history of man— "and there was a great earthquake, such as was not since men were upon the earth, so mighty an earthquake, and so great" (Rev. 16:18).

As the earth staggers in its orbit, the great cities of the world will crumble to ruins— "the cities of the nations fell" (Rev. 16:19). This will bring about the "full end of all nations" (D&C 87:6).[87] In their place will rise the world-wide Kingdom of God. It will be in Zion and her stakes that the saints will "stand in holy places" while being prepared in all things—temporally and spiritually, in principle and ordinance—for the day when the Son of Man comes in his glory with all of the holy angels to establish his kingdom and rule as King and Lawgiver upon his footstool for a thousand years.[88]

NOTES

1 Orson F. Whitney, *Saturday Night Thoughts,* 9-13; see JD, 26:200.
2 See D&C 84:110; Rev. 10:6. Time as such, being a natural extension of matter and space, never ends; it is eternal (see TPJS, 371).
3 To "remain in Babylon" is to be classified a telestial spirit. Such characters will not survive the coming of Christ in his glory. Virtually every prophecy concerning the end of the world is categorical in nature.
4 Most of mankind will eventually repent, be freed of Satan's influence, and be saved in the kingdom of God (see D&C 76:40-44).
5 Recall Paul's words to the rebellious saints in Corinth: "What will ye? Shall I come unto you with a rod, or in love, and in the spirit of meekness?" (1 Cor. 4:21).
6 While the Spirit is being withdrawn from the nations, it will not depart from righteous individuals (see TPJS,. 231).
7 A number of articles and books have been published on the "Gaia hypothesis" by a group of earth scientists who maintain that the earth is not a neutral, impassive entity, but a living body with the equivalence of senses, intelligence, and the capacity to act. "If men continue to antagonize it, the earth will become the most dangerous opponent ever to face the human race." However, the Creator controls his creations; the judgments of the last days are controlled judgments.
8 See D&C 87:6; 88:90.
9 See D&C 109:30; Rev.16:21.
10 See D&C, 29:13, 16; 43:18-25; 45:33, 41; 87:6, 88:89-91; 2 Ne. 6:15; 27:2; Morm. 8:29-30.
11 See D&C, 84:97; 87:6; Rev. 15; 16; 18; JST Matt. 24:30.
12 This delay is from our perspective; the Lord foreknew Zion would not be established by Joseph Smith. (see D&C 3:1-3; 58:3-7).
13 See TPJS, 17-19, 71, 160-61, 231-32
14 See D&C 105:1-6.
15 The times of the Gentiles is that period when the gospel is being preached to the nations prior to the redemption of Judah. When the

Lord's servants are called home (D&C 88:88), those times end; see JD, 2:261-62.
16. D&C 45:31-33; see 5:19; 97:23; Rev. 9:6.
17. Elder Bruce R. McConkie spoke of "the atomic holocausts that surely shall be" (CR, 1 Apr. 1979, 133).
18. Wilford Woodruff Journal, 7:419-23 (15 June 1878). Entitled, "A Vision," the account is introduced with the statement: "I had a very strange vision Copied in the office to day of a desolating sickness which Covered the whole land." He dates the vision 16 Dec. 1877. It has been variously attributed to Wilford Woodruff, John Taylor, Joseph F. Smith, etc. The author was reading the scriptures in French at the time the vision occurred. John Taylor spoke French and Wilford Woodruff began the study of French in November 1845. He is the most likely author.
19. See D&C 84:114-115.
20. See JD, 2:147; 5:218; 8:324;10:255; 12:119; 20:318; 20:151; 21:301.
21. D&C 101:18; see vs. 75; 103:11.
22. D&C 9:18-20; see Zech. 14:12; see JD, 14:351; 21:325.
23. CR, April, 1950, 5, 9, 169.
24. See D&C 1:7, 37-38; 101:64; 109:44-47.
25. *Messenger and Advocate,* 2: (Oct. 1835) 199; emphasis added. (Excerpt from account of early history of Joseph Smith by Oliver Cowdery). See Isa. 59:20. Orson Pratt also said the Ten Tribes would come to America after the latter-day Zion had been built and before the Second coming (see JD, 18:22-25, 68; 16:325).
26. Said Jehovah, "I am a father to Israel, and Ephraim is my firstborn" (Jer. 31:9).
27. Joseph Smith may be the very one who will turn the key for the Lost Tribes and be an active participant in the mighty events associated with their exodus to Zion. See *Autobiography of Parley P. Pratt,* 333; D&C 91:32;103:35.
28. See *The Way To Perfection,* 130.
29. See Abr. 2:8-11; Gen. 12:1-3.
30. D&C 7:3; see John 21:20-23. A similar request was made by three of the Nephite disciples who also became translated beings (see 3 Ne. 28:4-10).
31. The powers of translated beings are defined by Christ himself in 3 Ne. 28:7-9.
32. See Rev. 7:2-4; D&C 77:9, 14.
33. See Rev. 10. Note that the Savior's summary of John's labors in D&C 7:3 are restated in Rev. 10:11.
34. Shalmaneser V began the siege of Samaria, but died before its fall. His successor, Sargon II, completed the leading away of the Lost Tribes.

35 Minutes of the School of the Prophets (17 Aug. 1872); Parowan, Utah, 156-57.
36 See *The Way To Perfection,* 127-128,131; *Doctrines of Salvation,* 3:253.
37 Numbers 1:46 states that the twelve tribes consisted of more than six hundred thousand males alone.
38 See 23:7-8; 31:7-8; TPJS, 93.
39 See Jer. 16:16; 3:14.
40 This is the only scriptural reference associating the Lost Tribes with ice (see JD, 4:231; 18:24, 38, 127; 21:300-01; CR, Oct. 1913, 69; Oct. 1920, 32).
41 B.H. Roberts questioned the literalness of this passage: "Perhaps we have not always attributed sufficient importance to the imagery of poetry and revelation in giving interpretation to these scenes" (CR, Oct. 1928, 88). Bruce R. McConkie agreed: "In the literal sense of the word, the Ten Tribes will not return. . . with the ice flowing down at their presence; on a highway spanning oceans and continents. . . .They will tread the highway of righteousness. . .the wicked will have been destroyed and the Lord himself will be reigning on the earth. The return of the Ten Tribes is, of course, a Millennial event" (*A New Witness for the Articles of Faith*, 520-21).
42 See Isa. 43:6; 49:12; Jer. 3:18; 6:15; 23:8; 31:8; D&C 110:11; Ether 13:11.
43 The Prophet added, "But Judah shall obtain deliverance at Jerusalem."
44 See JD, 2:201. The only true prophets in the known world are those associated with The Church of Jesus Christ of Latter-day Saints. Those prophets ministering to the Lost Tribes will become subject to their keys and authority (see D&C 133:32).
45 See JD, 18:38, 127. Heber C. Kimball also spoke of their present prophets (see JD, 8:107). The imminent fulfillment of latter-day prophecies characterized LDS sermonizing from the days of Joseph Smith until recent times. As before stated, the prophets have been given God's agenda, but not his time table for carrying it out. We must remember that from his perspective, only a little over four hours have passed since the Church was established.
46 See JD, 2:201; 18:25, 38; 4:231-32;.
47 There is no unanimity of opinion on this point. Elder Bruce R. McConkie wrote of "*The myth of the Ten Tribes returning as guided by their prophets*" and denied that their return would literally involve "the ice flowing down at their presence; on a highway spanning oceans and continents," etc. (See *A New Witness for the Articles of Faith*, 520-21). For comments of earlier Church leaders on the subject see CR, Oct. 1913, 69; Oct. 1920, 32; April 1928, 57; April 1953, 83 and *Collected Discourses* 2:177; 3:30, 368; 5:38.

48 It has been theorized that the heat created by the friction produced when the earth's missing fragments are restored will be a factor in melting the frozen wastes of the Arctic (see Robert Smith, *The Last Days,* 79-80).
49 See Isa. 11:16.
50 Just how this "highway" will be formed is unrevealed. It may consist of the land mass on which the tribes were originally transported, or be formed from one rising out of the Arctic Ocean.
51 *Autobiography of Wandle Mace,* 35. Daniel Allen reported hearing Joseph Smith say that the return of that portion of the earth inhabited by the Lost Tribes would cause "a great shake" (Minutes of "School of the Prophets," Parowan, Utah, 17 Aug. 1872, 157; see D&C 88:87).
52 Isaiah also compared the planet to a frightened animal: "Therefore I will shake the heavens, and the earth shall remove out of her place. . . . And it shall be as the chased roe, and as a sheep that no man taketh up" (Isa. 13:13-14).
53 Journal of Samuel Holister Rogers, type copy, Brigham Young University, H. B. Lee Library, 17.
54 If the statement of Wandle Mace proves correct, the earth will reel in its orbit at least twice, first in connection with the returning Ten Tribes, and then with the returning city of Enoch.
55 The fact that the tribes will be confronted with enemies clearly indicates that they will return in a turbulent time prior to Christ's world coming since any possible "enemies" will have been destroyed by that consuming event (see JD, 10:26).
56 See TPJS, 161; JD, 17:4; 21:301. "This generation" may refer to a prevailing moral condition or state of affairs as well as to those born within a given time frame. See, for example, Matt. 16:4; Luke 21:32. The import of this prophecy is that the judgments described will occur in the time frame of the redemption of Zion and the subsequent return of the Lost Tribes (see JD, 18:22-23, 67-88).
57 The resurrected Christ quoted this prophecy to the Nephites in the context of the fate of latter-day Gentiles in America (see 3 Ne. 22:3). The neutron bomb is capable of destroying life without damaging property.
58 See Alma 45:16; Ether 2:7-12; 8:22-25.
59 CR, Oct. 1910, 76; see April 1916, 130.
60 See Mark 13:32-33.
61 See 2 Ne. 29:13; JD, 3:186; 9:212; 19:172; 21:301; 22:341-42.
62 Issued 6 April 1845 in New York City.
63 This point is reflected in a number of statements by Church leaders. See JD, 15:110; 23:300; CR, April 1916, 130; October 1956, 24; April 1963, 67.
64 It is not necessary for literally all those constituting the Lost Tribes

come to the center place of Zion, the New Jerusalem, in Missouri. They "come to Zion" when they enter the House of the Lord wherever they may be.

65 See Gen. 37:5-10.
66 The returning city of Enoch may well be "the sign of the Son of Man" foretold by Jesus (see Matt. 24:29-30).
67 See D&C 76:66-67; TPJS, 12.
68 See D&C 133:9; 82:14; 101:21; 109:59.
69 See D&C 88:93; 49:23.
70 See Joseph Smith Papers, Church Archives.
71 See D&C 36:8; 133:2; Mal. 3:1; TPJS, 340-41.
72 See Moses 7;16; D&C 84:101; 97:19; 3 Ne. 20:22; 21:22-23.
73 See also D&C 88:95; JD 8:51. The seventh seal mentioned in the book of Revelation pertains to events which are to transpire during the seventh thousand year period of the earth's temporal existence (see D&C 77:6-7). The "half hour" must refer to celestial, not earth, time—a period of about twenty-one years (see Abr. 3:4; 5:13; 2 Pet. 3:8). Prophecies of an eschatological nature are almost invariably couched in the language and time frame of God, not of man (see D&C 63:53).
74 See D&C 86:4-7;101:65-66. This is the period in which the 144,000 will labor in sealing up the worthy into the Church of the Firstborn. See D&C 77:11; Rev. 7:3-8; WJS, 297, n. 7.
75 See Rev. 8, 9, 16.
76 See D&C 29:14; 45:9-10.
77 The Lord is not cruel; his judgments are swift and brief. The absence of the rainbow will signify a worldwide drought among the rebellious. It may afflict the righteous in Zion to some extent, but this is questionable (see Moses 8:4, 9).
78 See Joel 2:10, 31; 3:15; Acts 2:20. Zion will be redeemed and the New Jerusalem established before the return of the Lost Tribes and the selection of the 144,000, which events will occur in the closing period of the sixth seal (see Rev. 7:1-4; 8:1; D&C 77:9-11). The deliverance of Judah will occur after the opening of the seventh seal and during the final phase of the judgments which will immediately precede Christ's world coming.
79 See Mark 13:38; Luke 21:25.
80 See Rev. 6:12-13; D&C 29:14; 34:9; 45:42; 88:87. In Revelation, the sun is described as becoming "black as sackcloth of hair;" in the Doctrine and Covenants as being "darkened" or as hiding "his face." In Revelation, the moon is described "as blood;" the Doctrine and Covenants states that the moon will be turned into blood" or "bathed in blood" or, simply, "withhold its light."
81 See D&C 88:87; Isa. 34:4.

82 In scripture, the star is a symbol for man. The morning stars who sang together were the pre-mortal sons of God (Job 38:7 and D&C 128:23), Lucifer, a "son of the morning" sought to exalt himself above "the stars of God" (Isaiah 14:12-14), he fell as a star to earth (Revelation 9:1) and a third part of the stars with him (Revelation 8:12; 12:4, 8-9). Abraham, Isaac and Jacob were promised that their posterity would be as the "stars of heaven" (Genesis 15:5; 22:17; 26:4; Exodus 32:13; Deuteronomy 28:62; 1 Chronicles 27:23; D&C 132:30). Joseph's brothers were the eleven stars who bowed to him (Genesis 39:9-10). Daniel saw the "little horn" oppress the stars, the Lord's people (Daniel 8:10). Stars symbolized the seven servants (angels) and twelve apostles (Revelation 1:20; 3:1; 4:5; JST Revelation 5:6; 12:1 and 1 Nephi 1:10). Christ is the "star out of Jacob" (Numbers 24:17) and the "bright and morning star" (Revelation 2:28; 22:16; 2 Peter 1:19).
83 Indeed, John's very language suggests that actual stars will be hurled through space as glorious heralds of the Savior's second advent.
84 In Scripture, the star is a symbol for man. The morning stars who sang together were the pre-mortal sons of God (Job 38:7; D&C 128:23), Lucifer, a "son of the morning" sought to exalt himself above "the stars of God" (Isa. 14:12-14), he fell as a star to earth (Rev. 9:1) and a third part of the stars with him (Rev. 8:12; 12:4, 8-9). Abraham, Isaac and Jacob were promised that their posterity would be as the "stars of heaven" (Gen. 15:5; 22:17; 26:4; Ex. 32:13; Deut. 28:62; 1 Chron. 27:23; D&C 132:30). Joseph's brothers were the eleven stars who bowed to him (Gen. 39:9-10). Daniel saw the "little horn" oppress the stars, the Lord's people (Dan. 8:10). Stars symbolized the seven servants (angels) and twelve apostles (Rev. 1:20; 3:1; 4:5; JST Rev. 5:6; 12:1 and 1 Ne. 1:10). Christ is the "star out of Jacob" (Num. 24:17) and the "bright and morning star" (Rev. 2:28; 22:16; 2 Pet. 1:19).
85 See also D&C 45:48; 49:23; Isa. 13:13-14; 24:18-20. Joseph Smith also put the reeling of the earth in the context of stars falling when he said that "the time is near when the sun will be darkened , and the moon turn to blood, and the stars fall from heaven, and the earth reel to and fro" (TPJS, 71). It appears that the planet will "reel" to some extent at least three times: when the Lost Tribes return, when Christ appears to the Jews, and when Enoch's city returns.
86 See D&C, 84:118; Zech. 12-14; Rev. 16:19.
87 See Rev. 6:12-13; 11:13; 16:14-21. For the brief period then remaining before Christ's world advent, peoples will cluster together in loosely organized tribal arrangements (see JST Matt. 24:37). A similar condition existed among the Nephites following the fall of their republic just prior to Jesus' crucifixion (see 3 Ne. 7:1-6).
88 See D&C 45:59; 38:22; 41:4; 43:29; 58:22

Chapter Eleven

Earth's Day of Rest

And I have made the earth rich, and behold it is my footstool, wherefore again I will stand upon it (D&C 38:18).[1]

The return of Jesus Christ is more certain than the rising of the sun tomorrow morning—for he swore with an oath he would come again, even as he swore with an oath Earth would never suffer a second global Flood.

> And the Lord said unto Enoch: *As I live,* even so will I come in the last days, in the days of wickedness and vengeance, to fulfill the oath which I have made unto you concerning the children of Noah (Moses 7:60).

Christ promised Enoch he would "call upon the children of Noah" (Moses 7:51). That "call" is his second coming.

At the time of his crucifixion, Earth cried out, "When will my Creator sanctify me, that I may rest, and righteousness for a season abide upon my face?" (Moses 7:48).[2] Enoch, in vision, heard that cry and asked: "When the Son of Man cometh in the flesh, shall the earth rest?" (Moses 7:54). No, there would be no rest for Mother Earth in the first coming of the Son of Man. To the contrary, it would be a time of great turmoil when all creation mourned the death of God. Earth would have to wait—together with all those holy men and women who had sought that day of righteousness without ever finding it. These ancient saints "confessed they were strangers and pilgrims on the earth; but obtained a promise that they should find it and see it in their flesh" (D&C 45:12-13).³ That promise will be kept on Earth's second Sabbath in the seventh millennium after the Fall. Its dark, pre-dawn hours will vanish before the brilliance of its Day Star—the Son of Man. Only the worthy will abide his appearing; only they will be permitted to keep "the Sabbath of creation" (TPJS, 103).

Christ Comes in Glory

The Great Unveiling

In being cast down to this benighted corner of the universe, Earth, like Adam, passed through a spiritual veil into a telestial state. This passage marked the beginning of its temporal existence. The Father's children pass through a similar spiritual veil when they come to Earth and enter their fallen physical bodies. Indeed, the mortal body constitutes *a second veil* between them and their divine Parent.[4]

Jesus, too, accepted the limitations imposed by the Fall: "Forasmuch then as the children are partakers of flesh and blood, he also himself likewise took part of the same. . . . For verily he took not on him the nature of angels; but he took on him the seed of Abraham" (Heb. 2:14, 16).[5] Christ in the flesh was Christ veiled.[6] His immortal glory was hidden behind the common facade of humanity. Mankind looked upon the face of God, but did not recognize him.

How different his return! God in his glory is a "consuming fire" (Deut. 4:24). and who will "abide the day of his coming? and who shall stand when he appeareth? for he is like a refiner's fire, and like fuller's soap" (Mal. 3:2; see D&C 128:24). The answer: only the sanctified and those who have been quickened by a portion of terrestrial glory will be able to endure his presence.[7] Unlike his obscure entrance into this world as the babe of Bethlehem, the Savior's return will be both public and universal:

> For as the light of the morning cometh out of the east, and shineth even unto the west, and covereth the whole earth; so shall also the coming of the Son of Man be (JST, Matt. 24:27).

The saints are commanded to watch for that awesome revelation "when the veil of the covering of my temple, in my tabernacle which *hideth the earth,* shall be taken off, and *all flesh* shall see me together" (D&C 101:23).[8] Earth's Lord will pass through that veil and burst upon an unprepared and astonished world. He will descend from the celestial heavens— "the regions which are not known, clothed in his glorious apparel, traveling in the greatness of his strength"—to claim his rightful throne. The "curtain of Heaven" will part "and the face of the Lord shall be unveiled" (D&C 88:95). And when it is, the heavens will be bathed in a divine radiance of such transcendent splendor as to swallow up all lesser lights. As God possesses more light and truth than the combined intelligences over which he reigns,[9] so will the glory emanating from the presence of Jesus Christ overwhelm and blot out the combined brilliance of sun, moon and stars:

> And so great shall be the glory of his presence that the sun shall hide his face in shame and the moon shall withhold its light, and the stars shall be hurled from their places (D&C 133:49).[10]

Earth's Spiritual Baptism

We have seen that the Fall which alienated man from God, also alienated the Earth from its Creator. The *ordinance of reconciliation* is the same for both—baptism. Not the partial baptism of water, but the *total* baptism of water, fire, and the Holy Ghost.[11] Earth was "born of water" when it was immersed in the Flood; it will be born "of the Spirit" when it is immersed, first in fire, and then in the Holy Ghost. Brigham Young said:

> This earth, in its present condition and situation, is not a fit habitation for the sanctified; but it abides the law of its creation, has been baptized with water, will be baptized by fire and the Holy Ghost, and by-and-by will be prepared for the faithful to dwell upon (JD, 8:83; see 26:266).

Christ's coming in glory and the baptism of the earth in fire and the Holy Ghost occur at one and the same time.

The eternal law of affinities assures like can only cleave to like. A holy, incorrupt God cannot dwell in glory on an unholy, corrupt earth or among a corrupt people. Hence the necessity of a literal refining fire which will burn away all that is defiling and unclean. Orson Pratt commented:

> It must be cleansed by an element that is stronger and more purifying than that of water, namely, the element of fire. Fire must prevail over all the face of this earth. . . .So this earth in due time must be baptized with fire first, and then the Holy Ghost. Fire will cleanse all the proud and they that do wickedly from its face—all persons that are corrupt, all sinful persons, all disobedient persons, all who do not keep the commandments of God; it will cleanse the earth by burning them as stubble, fulfilling the words of the prophet Malachi (JD, 21:324; see1:331; 16:319).

The very elements comprising the face of the earth, together with every work of man by which he has polluted it physically or morally will come under this fiery judgment; nothing will escape; everything will be weighed in the Lord's balances. The Apostle Peter wrote:

> The heavens shall pass away with a great noise, and the elements shall melt with fervent heat, the earth also *and the works that are therein* shall be burned up (2 Pet. 3:10).

In quoting Malachi's prophecy to young Joseph Smith—"For behold, the day cometh, that shall burn as an oven, and all the proud, yea, and all that do wickedly shall burn as stubble"—the angel Moroni added a most significant clause: *"for they that come shall burn them"* (JS-H 1:37; emphasis in original). The world will not end in a nuclear holocaust, but in the judgments of a holy God. It will not pass away in darkness, but in the purifying presence of the Son of Man and that vast concourse of resurrected beings who will accompany him. The intense spiritual energy—the glory—emanating from the Son and his celestial companions shall be all-consuming: "For *the presence of the Lord* shall be as the melting fire that burneth, and as the fire which causeth the waters to boil" (D&C 133:41).[12] The contaminated surface of this planet will be consumed to nothingness. Every form of life—whether man, animal, or vegetation; whether of land, sea or air—of a corrupt, defiled nature will be *selectively* consumed[13] in the day that "shall burn as an oven."[14] Not even those inferior forms of marine life which lived through the Flood will escape the global firestorm. The very seas will boil so that only those creatures fit to live in terrestrial waters will survive.[15] The Lord declared:

> And every corruptible thing, both of man or of the beasts of the field, or of the fowls of the heavens, or of the fish of the sea, that dwells upon all the face of the earth, shall be consumed; And also that of element shall melt with fervent heat; and all things shall become new, that my knowledge and glory may dwell upon all the earth (D&C 101:24-25).[16]

Newly born out of fire, Earth will be cleansed of all of the unholy works of man, great and small—virtually nothing will remain of six thousand years of human agency and effort.

Nothing, that is, except those things written upon the hearts and minds of all who had lived in the previous six thousand years. Such is the glory and the vanity of this world.

A Paradisiacal Heaven and Earth

> For, behold, I create new heavens and a new earth: and the former shall not be remembered, nor come into mind (Isa. 65:17).[17]

Joseph Smith confirmed Isaiah's testimony: "the earth will be renewed and receive its paradisiacal glory" (10th Article of Faith). With the completion of the total ordinance of baptism, Earth, having been cleansed by fire and sanctified by the enlivening power of the Holy Ghost unto the renewing of its body,[18] will be a truly "born again" planet—born back into its Edenic state and into the literal presence of God the Son who, as Son, is the God of terrestrial glory.[19]

The enveloping radiance of the paradisiacal Earth will far exceed that of its former telestial state. According to Isaiah, the millennial moon will equal the brilliance of the present Sun, and the Sun will, in turn, be sevenfold brighter than it is now.[20] Only those living things which have been quickened by terrestrial powers will be able to abide its millennial environment. Like Nebuchadnezzar's fiery furnace (which presaged it), the earth will be heated "seven times more than it was wont to be heated." Being so, it would consume things telestial just as the flames of Nebuchadnezzar's fires consumed those who sought to destroy the Shadrach, Meshach, and Abednego.[21] But Earth's paradisiacal inhabitants will be bathed in the enlivening glory of the fourth man, the Son of God. He will be in the midst of his people even as he stood with the three Hebrews in that fiery furnace in ancient Babylon!

All this that Earth might begin its foreordained journey back into the celestial presence of the Father whose hidden mystery was revealed by Paul:

> That in the dispensation of the fulness of times he might gather together in one all things in Christ, both which are in heaven, and which are on earth; even in him (Eph. 1:10).

Christ, the Eternal Reconciler, will make of many, one, restoring all things to that harmonious order which prevailed in the beginning. All division will end, all alienation will cease; perfect unity will prevail throughout the Earth in consequence of the restoration of all things both physical and spiritual.[22]

THE RESTORATION OF ALL THINGS

"The works, and the designs, and the purposes of God cannot be frustrated, neither can they come to naught" (D&C 3:1). Whereas the agency of man may disrupt the divine order, the agency of God ultimately restores it. The restoration of all things began with the labors of the Prophet Joseph Smith. Through him was restored lost scriptures, the fullness of the gospel, the keys and authorities of the Holy Priesthood, the true and living Church of Jesus Christ, saving and exalting ordinances, and all else once had among men. This restoration of things spiritual will be accompanied by the restoration of things temporal. In the future we can anticipate the redemption of the land of Zion in Missouri and the building of the New Jerusalem. Thereafter, the Lost Tribes will return and, in due time, be restored to their lands of promise. The planet's continents will be restored to their pre-division state. The "great deep" will, in turn, assume its former relationship to the land. In fine, all things physical will be restored to the paradisiacal state existing in the beginning of time.

Reunion of the Continents

The Lord is a perfectionist, he will not be satisfied until every jot and tittle of prophecy has been fulfilled.[23] When he ministered to Israel in America, Mormon tells us:

> And he did expound all things, even from the beginning until the time that he should come in his glory—yea, even all things which should come upon the face of the earth, even until the elements should melt with fervent heat, and the earth should be wrapt together as a scroll, and the heavens and the earth should pass away (3 Ne. 26:3).[24]

Mormon asks the Gentiles of our day: "Know ye not that he hath all power, and at his great command the *earth shall be rolled together* as a scroll?" (Morm. 5:23).[25] That is, the continents will be brought together even as the writings on a scroll are brought together when it is rolled up. The division of the planet's original land mass into the various continents during the lifetime of Peleg will be reversed in connection with the restoration of all things. In 1831, the Lord spoke in plainness of this soon-to-be phenomenon:

> He shall *command* the great deep, and it shall be driven back into the north countries, and the islands shall become *one land;* And the land of Jerusalem and the land of Zion shall be turned back into their own place, and the *earth shall be like as it was* in the days before it was *divided.* (D&C 133:23-24).

Joseph Smith restated this prophecy in 1844:

> . . .yea, the Eternal God hath declared that the great deep shall roll back into the north countries and that the land of Zion and the land of Jerusalem shall be joined together, as they were before they were divided in the days of Peleg.[26]

Orson Pratt accepted the literalness of this amazing prophecy:

> The earth is now divided into continents and islands. We may ask, are these to change their location? The answer is, yes. . . .Such islands as Great Britain will change their location, as well as those of the Pacific Ocean and all others in like manner; and I have no doubt there will be a vast change between the location of the great oceans and seas at that time. The earth will doubtless be rolled back to the position it formerly occupied (JD, 18:315).[27]

The wedding of the land of Jerusalem (Judah) and the land of Zion (Joseph) will fulfill Isaiah's beautiful metaphor concerning those cities from which the law of the Lord shall issue forth:

> Thou shalt also be a crown of glory
> In the hand of the Lord,
> And a royal diadem in the hand of thy God.
> Thou shalt no more be termed Forsaken;
> Neither shall thy land any more be termed Desolate:
> But thou shalt be called Hephzibah,[28]
> And thy land Beulah:[29]
> For the Lord delighteth in thee,
> *And thy land shall be married* (Isa. 62:3-4).[30]

The "marriage" of the continents will be another of Christ's unprecedented acts in the time of the seventh seal. Scripture indicates they will be reunited in connection with his Jerusalem appearance to a remnant of Judah. This will be the occassion, according to John the Revelator, of a worldwide earthquake when the planet "reels to and fro" and when *"every island fled away,* and the *mountains were not found"* (Rev. 16:20). Earth's land masses at the time of its spiritual baptism will be as they were when it was baptized in water: the continents were united in the beginning; they will be united in the end. Earth will have come full circle. Just what the planet's configuration will be in that day is unrevealed.[31] What matters is that "the earth shall be like as it was in the days before it was divided."

The Great Deep

When "the great deep" is "driven back into the north countries," the natural consequence will be "the islands shall become one land." (D&C 133:23).[32] The Pacific Ocean alone averages fourteen thousand feet in depth and covers some seventy million square miles, or one third of the planet's surface. It contains about twenty thousand islands.[33] These islands would be absorbed into that vast area—with its present undersea valleys and mountains—now known as the Pacific basin.

Where will all the waters of "the great deep" go? Some surface waters may be channeled back into the bowels of the planet— "the fountains of the great deep" which were "broken up" in connection with the Flood.[34] And the restoration of the planet to its ante-diluvian condition would involve those waters which existed "above the expanse" beginning on the second day of Creation.[35] Are those waters to be restored to their former location? If so, they might serve as a canopy against the harmful cosmic radiation now bombarding the planet and contribute to the prophesied future longevity of mankind.[36] However, such radiation may not be a factor when the planet is enveloped in paradisiacal glory. Then too, if the aforementioned testimonies regarding the planet's original size prove correct, it will be able to accommodate the "great deep" wherever it is driven.

Even without the displacement of the planet's present seas and oceans, a melting of polar ice fields would inundate the coastal areas of the present continents. However, the elevations of these regions doubtlessly will be modified when the continents are rejoined by the Lord who will also "break down the mountains" so that "the valleys shall not be found" (D&C 133:22). But we need not be overly concerned with the logistics of prophecy. If God knew enough to create this planet, he probably knows enough to make any necessary modifications to it in the future.

The Mountains are Humbled

When the whole planet becomes the mountain of the Lord—the dwelling place of God—there will be no need for any other high places. The Lord will be everywhere: "The earth shall be full of the knowledge of the Lord, as the waters cover the sea" (Isa. 11:9; see Hab. 2:14). This knowledge will be personal not vicarious because, said the Lord, "all shall know me, who remain, even from the least unto the greatest, and shall be filled with the knowledge of the Lord, and shall see eye to eye" (D&C 84:98; see Jer. 31:34).

The temporal and spiritual equality which will characterize millennial society will have its counterpart in the planet's very topography. The great mountain ranges will be no more. They, too, are symbols of this fallen world and will be brought low before their Creator even as men are to be humbled before him:

> And the loftiness of man shall be bowed down, and the haughtiness of men shall be made low: and the Lord alone shall be exalted in that day (Isa. 2:17; see D&C 104:16).

Isaiah's prophecy became John's message: "Every valley shall be exalted, and every mountain and hill shall be made low: and the crooked shall be made straight, and the rough places plain" (Isa.40:4).[37] Joseph Smith wrote that the ancient prophets and seers:

> ...saw the glory of the Lord when he showed the transfiguration of the earth on the mount; they saw every mountain laid low and every valley exalted (TPJS, 13).

Looking to that day, the psalmist wrote: "The hills melted like wax *at the presence of the Lord*, at the presence of the Lord of the whole earth" (Ps. 97:5). His fiery glory will envelop the earth and the mountains will become molten and flow down until they become one with the valleys.[38] This, together with the planet-wide shifting of the ocean basins and continental land masses, will result in a millennial earth characterized by lesser mountains, rolling hills and broad valleys. As it was in the beginning, so will it be in the time of the restoration of all things. The physical barriers which the Lord established after the Flood to divide the nations will be no more. Both spiritual and physical unity will prevail.

Magnitude of the Millennial-Earth

While neither the original nor the final dimensions of Earth are known, a few interesting comments on the subject have been made. The substance of them all is that Earth was considerably larger when first organized and that, consequently, it will be considerably larger hereafter. In the winter of 1840-41 Samuel Holister Rogers attended a public meeting held at Vincent Knight's home during which the Prophet Joseph Smith gave the following instruction:

> When this world was first made it was a tremendous big thing. The Lord concluded it was to[o] big. We read in the Scriptures that in the days of Peleg the earth was divided so the Lord divided the earth. When the ten tribes of the children of Israel went into the north country he divided it again, so the earth has been divided and subdivided.[39]

Eliza R. Snow confirms the essence of Rogers' journal entry in the following stanzas of her poem, "Address to Earth":

> But thy dimensions have been torn
> Asunder, piece by piece
> And each dismember'd fragment borne
> Abroad to distant space.
>
> **************
>
> And thus, from time to time, thy size
> Has been diminished, till
> Thou seemst the law of sacrifice
> Created to fulfill.
>
> **************
>
> A 'restitution' yet must come,
> That will to thee restore,
> By the grand law of worlds, thy sum
> Of matter heretofore.[40]

Parley P. Pratt simply wrote, "The earth will be much larger than it is now" (*MS*, 1: [Feb. 1841] 258).

Earth is Healed

We have seen that the fall of Adam together with the subsequent sins of his children down through the generations led to the Earth being repeatedly cursed against mankind. However, the great Jehovah will come with "healing in his wings" in fulfillment of his ancient promise to "heal the land" when his people turned from their sinful ways.[41] Brigham Young felt that the Latter-day Saints would play a part in Earth's redemption:

> The curse which has been brought upon the earth through the Fall will be removed through the faith and virtues of the Saints. When we become sanctified in the truth, and our faith, through the Gospel of the Son of God, becomes suffi-

ciently powerful we will be able to remove the thorns and thistles and obnoxious weeds that grow immediately around us, and to bless and sanctify our gardens and farms, so that they will bring forth spontaneously the fruits and flowers, the cereals and vegetables that sustain life; and upon this principle as righteousness extends will the whole earth eventually be redeemed and sanctified, when all things will be as they were in the beginning, when the Lord finished the earth and pronounced everything to be 'very good' (JD, 19:4).[42]

The Lord has promised the righteous who dwell in the Zion "the good things of the earth, and it shall bring forth in its strength" (D&C 59:3).

In that day shall the branch of the Lord be beautiful and glorious, and the fruit of the earth shall be excellent and comely for them that are escaped of Israel (Isa. 4:2)

For the Lord shall comfort Zion: he will comfort all her waste places; and he will make her wilderness like Eden, and her desert like the garden of the Lord; joy and gladness shall be found therein, thanksgiving, and the voice of melody (Isa. 51:3).

Being both the pattern and the nucleus of terrestrial life in the millennium—the united Zions of Joseph, Enoch and Judah will be harbingers of better days for all mankind when the Kingdom of God is established in its millennial fullness.[43] The blessings bestowed upon these cities will be magnified, becoming the blessings of all mankind. And the peace, prosperity, and happiness of the relatively few will become the common lot of every nation, kindred, tongue, and people. No longer will Earth be burdened with sterile deserts and wastelands; the "river of God" will flow generously (see Ps. 65:9).

And in the barren deserts there shall come forth pools of living water; and the parched ground shall no longer be a thirsty land (D&C 133:29).

These words reflect those of Isaiah:

I will open rivers in high places, and fountains in the midst of the valleys: I will make the wilderness a pool of water, and the dry land springs of water (Isa. 41:18; see 35: 6-7).

He further wrote the "trees of the field shall clap their hands. Instead of the thorn shall come up the fir tree, and instead of the brier shall come up the myrtle tree" (Isa. 55:12-13; see JD, 24:210). Earth's longed-for season of rest will be one long harvest time:

Behold, the days come, saith the Lord, that the plowman shall overtake the reaper, and the treader of grapes him that soweth seed; and the mountains shall drop sweet wine, and all the hills shall melt (Amos 9:13).

This blessing will extend to peoples everywhere. Like a beloved wife and mother, Earth will yield of her strength—gladly, lavishly, continually:

The earth hath travailed and brought forth her strength;
And truth is established in her bowels;
And the heavens have smiled upon her;
And she is clothed with the glory of her God.
For he stands in the midst of his people (D&C 84:101).[44]

The Father delights in endowing his children with those things which make for joy—soul happiness. His message is one of glad tidings, of the more abundant life. His purpose is not to render his children sterile of all that makes them human, but to sanctify and immortalize their humanity so that it becomes everlastingly divine. Only then can they become one with him. For he is not passionless; he is a sensate being. His life is rich with sanctified sensory pleasures. He told Joseph Smith:

> All things which come of the earth, in the season thereof, are made for the benefit and the use of man, both to please the eye and to gladden the heart. Yea, for food and for raiment, for taste and for smell, to strengthen the body and enliven the soul (D&C 59:18-19).[45]

Brigham Young affirmed life:

> Our senses, if properly educated, are channels of endless felicity to us. . . . Everything that is joyful, beautiful, glorious, comforting, consoling, lovely, pleasing to the eye, good to the taste, pleasant to the smell, and happifying in every respect is for the Saints. . . . Every faculty and power of both body and mind is a gift from God (JD, 9:244).

Universal Peace and Justice

The Son of Man, the Prince of Peace, will declare a thousand year armistice in the war which began in the courts of heaven. Satan will be driven from the earth; his kingdom consumed in the glory of God. The governments of men with their attendant corruption, ineptitude, and oppression will likewise crumble into dust. The image of Nebuchadnezzar's dream—like Shelley's "Ozymandias"—will lie in ruins.[46] In its place will stand the kingdom of God; Daniel's vision will be fulfilled.

> And he shall judge among the nations, and shall rebuke many people: and they shall beat their swords into plowshares, and their spears into pruning hooks: nation shall not lift up sword against nation, neither shall they learn war any more (Dan. 2:4).

There will be no need to "learn war anymore" when war's basic causes—inequality and unrighteous dominion—are done away. James wrote that "wars and fightings" stem from the lusts of men for things they do not have.[47] And the Lord said, "it is not given that one man should possess that which is above another, wherefore the world lies in sin." A world in sin is a world at war. Christ's righteous reign will correct all material imbalances so that the needs and wants of every soul will be met without respect of persons or their places of habitation. No longer will there be "have" and "have-not" nations. No longer will the few glut themselves on the earth's natural resources while the many suffer privation for the want of the merest necessities of life. When war is no more, and human energy is fully directed toward creation rather than destruction, Jesus' words, "ye have the poor always with you," will no longer apply:

> For they shall see the kingdom of God coming in power and great glory unto their deliverance; for the fatness of the earth shall be theirs. For behold, the Lord shall come, and his recompense shall be with him, and he shall reward every man, and the poor shall rejoice; And their generations shall inherit the earth from generation to generation, forever and ever (D&C 56:18-20; see Isa. 65:21-23).

The tempering of the climate, together with the union of the continents and the elimination of the great mountain ranges, deserts, and arid regions will transform the entire planet into one vast promised land. Paul's "middle wall of partition"[48] which divided Israel from the Gentiles will vanish away so that the physical and political barriers which now separate the children of God from one another will be no more. All righteous peoples—being "equal in earthly things"—will become "equal in the bonds of heavenly things."[49] For a season peace and happiness will prevail globally. The "great division" at Christ's coming will be followed by a great union during his reign.⁵⁰ Earth's King will say, "Peace, peace, be still" to every living thing, and universal harmony will envelop the earth. This will bring to pass Isaiah's transforming prophecy:

> The wolf also shall dwell with the lamb, and the leopard shall lie down with the kid; and the calf and the young lion and the fatling together; and a little child shall lead them. And the cow and the bear shall feed; their young ones shall lie down together; and the lion shall eat straw like the ox. And the sucking child shall play on the hole of the asp, and the weaned child shall put his hand on the cockatrice' den. They shall not hurt nor destroy in all my holy mountain; for the earth shall be full of the knowledge of the Lord, as the waters cover the sea (Isa. 11:6-9; see 65:25).⁵¹

Isaiah's words are confirmed in a modern revelation: "And in that day the enmity of man, and the enmity of beasts, yea, the enmity of all flesh, shall cease from before my face" (D&C 101:26).⁵² Since all forms of life will be mutually secure, all will be purged of their carnivorous appetites, becoming as they were in the beginning before the Fall. To this end, Hyrum Smith wrote:

> [God]...knows what course to pursue to restore mankind to their pristine excellency and primitive vigour, and health: and he has appointed the word of wisdom as one of the engines to bring about this thing, to remove the beastly appetites, the murderous disposition and the vitiated taste of man; to restore his body to health, and vigour, promote peace between him and the brute creation, and as one of the little wheels in God's designs, to help regulate the great machinery, which shall eventually revolutionize the earth, and bring about the restoration of all things, and when they are restored he will plant "the tree of life, whose leaves shall be for the healing of the nations."⁵³

Since, as Isaiah foretold, "They shall not hurt nor destroy in all my holy mountain," flesh will not be eaten by man or beast—all will be herbiverous once more. Mankind's food will consist primarily of milk and honey, fruits and vegetables, grains and nuts—the Lord's millennial "Word of Wisdom." The sicknesses and diseases, the trials and sufferings— physical, emotional, moral, and spiritual—which now afflict humanity will be things our millennial posterity will not know.

And the Holy Spirit which godless men have all but driven from the face of the earth in these last days will return in mighty power even as they depart. The spiritual voids created over many centuries by the prince of darkness will be filled with the terrestrial light of Christ. The very atmosphere will become so charged with the Spirit that all living things will breathe it like air. President Lorenzo Snow testified:

> The whole earth is the Lord's. The time will come when it will be translated and be filled with the spirit and power of God. The atmosphere around it will be the spirit of the Almighty. We will breathe that Spirit instead of the atmosphere we now breathe (MS, 61 [August, 1899], 546).

The Spirit is not only peace, it is also life. As it was in the beginning, so will it be at the end: both spiritual and temporal death will be no more and mankind's lost longevity will be somewhat restored.[54]

> In that day an infant shall not die until he is old; and his life shall be as the age of a tree; And when he dies he shall not sleep, that is to say in the earth, but shall be changed in the twinkling of an eye, and shall be caught up, and his rest shall be glorious (D&C 101:30-31).[55]

Mankind will also regain its lost excellence and enter upon a life far superior to our telestial lives in spirituality, intelligence and moral purity. The curse of Babel will be no more.

The myriad languages which have served to foment discord, division, and war among nations will yield to a common tongue:

> For then will I turn to the people *a pure language,* that they may call upon the name of the Lord, to serve him with one consent (Zeph. 3:9).

We noted that Adam had such a language. As part of the restitution of all things, a universal, undefiled language will be spoken. Necessarily so, for the cleansing of the Earth includes the cleansing of the hearts and minds of its inhabitants. If hearts and minds are cleansed and filled with the Spirit of the Lord, truth and virtue will abound, divisive and unclean thoughts have no soil in which to grow. And if there are no evil thoughts, there can be no evil deeds. This is a major reason children of millennial saints "shall grow up without sin unto salvation" (D&C 45:58).

Earth's Hour

In discussing the organization of this planet, Joseph Smith referred to "the worlds which were created at the time" (TPJS, 348-49). Christ is the Savior of those worlds as well as this one; it is through him that "the inhabitants thereof are begotten sons and daughters unto God" (D&C 76:24).[56] However, in becoming the foreordained setting for the atonement which encircles those worlds within the arms of divine mercy, Earth, like its Creator, has descended below all things. Consequently, as noted by Brigham Young, our earth is "capable of becoming one of the highest kingdoms that has ever had an exaltation in the eternities" (JD, 10:175). It is the keystone of this eternity: from it holy men such as Enoch have gone forth to serve the needs of peoples on other planets who, said Joseph Smith, "shall be heirs of salvation" (TPJS, 170). This inter-planetary ministry reflects the universal labors of Christ and harmonizes with President John Taylor's declaration that the Father, Son and Holy Ghost constitute the "First Presidency of this system and this eternity" (*MA*, 76).[57]

He further wrote that the "twelve kingdoms" mentioned in a parable in the Doctrine and Covenants[58] are under the direction of this "Presidency." In the parable, the Lord likens his kingdoms to "a man having a field" into which he sends twelve servants with the promise that he will visit each of them:

> . . .every man in his hour, and in his time, and in his season—Beginning at the first, and so on unto the last, and from the last unto the first, and from the first unto the last (D&C 88:58-59).[59]

It is significant that three cycles of visitation are mentioned with the third one ending, not with the first kingdom, but with the last. Since the principle is immutable that the last shall be first in all things God has created, this Earth, which "abideth the law of a celestial kingdom," may well be the "twelfth" kingdom of the parable—the last to be organized in this creative epoch, and the last to be visited by its Creator. Being the last which shall be first, it could, indeed, become "one of the highest kingdoms that ever had an exaltation in the eternities."

And this, the dispensation of the fulness of times, being the last of all, is, likewise, the greatest of all—being the sum of all previous dispensations. It then follows that Joseph Smith, the eternal president of this dispensation is one of the greatest servants of God to be found in any of his kingdoms.[60] In any event, Earth's time with her Creator will come in fulfillment of the parable of the twelve kingdoms. Jesus Christ will spend an "hour" with this earth and its millennial inhabitants so that he might be glorified in them, and they in him— "that they all might be glorified" (D&C 88:60).

The millennial Earth will enjoy a paradisiacal state of peace, progress, and abundance. More importantly, it will be the setting of the climactic spiritual labors of the Church of Jesus Christ for the posterity of Adam from the beginning to the end of time. When all is done, its temporal day will end and it will turn toward its ultimate destiny. Having been the footstool of God the Son, Earth is also destined to become the footstool of God the Father as well.[61] To do so, it must leave this solar system, and the mortal stars—those symbols of that lesser kingdom to which it fell after the transgression of Adam—and assume its foreordained place near the throne of God. However, before it can be infused with celestial glory it must pass through the veil of death. It must "flee away to make place for the city of God" (TPJS, 13).

NOTES

1 Several scritural passages describe Earth as the footstool of the Lord: Isa. 66:1; Matt. 5:35; Acts 7:49; Abr. 2:7; Moses 6:9, 44; 1 Ne. 17:39;. Israel's ancient tabernacle and temple were also referred to as the Lord's footstool. See 1 Chron. 28:2; Ps. 99:5; 132:7.
2 Note that divine "rest" follows sanctification. The Holy Ghost is the instrument of sanctification; see Alma 13:12..
3 See Heb. 11:13-16; 1 Pet. 2:11.
4 Brigham Young said: "Because we are encumbered with this flesh, we are in darkness; the flesh is the veil that is over the nations" (JD, 4:134; emphasis added).
5 Although Jesus' divine conception rendered him immortal, still he subjected himself to all of the natural vicissitudes of the flesh; see Heb. 4:15; Mosiah 3:7; Alma 7:11-12.
6 Jesus came to earth in his glory, a fullness of the Spirit; see 1 Ne. 11:28; Alma 13:24; TPJS, 187-88.
7 See D&C 35:21; 38:8; 45:57.
8 See 109:74; 124:8.
9 See Abr. 3:19-21; TPJS, 353, n.8.
10 This phenomenon should not be confused with the latter-day obscuring of the heavenly bodies in consequence of the global upheavals which will fill the atmosphere with thick dust (see 3 Ne. 8:19-22).
11 Joseph Smith stated: "Baptism by water is but half a baptism, and is good for nothing without the other half—that is, the baptism of the Holy Ghost" (TPJS, 314).
12 See Isa. 64:2; Job 41:31.
13 Just as Daniel's three Hebrew companions survived the fiery furnace of Nebuchadnezzar, so will the Son of God preserve all worthy forms of life in that day when the earth "shall burn as an oven" (see Dan. 3).
14 See Mal. 4:1; D&C 64:24; TPJS, 13.
15 See D&C 133:41; Isa. 64:1-2.
16 See 3 Ne. 26:3; Morm. 9:2; Isa. 34:4. The destruction of all extant telestial life forms prior to the earth's renewal is probably the final phase of a process of cleansing which began after the Fall and was accelerated by the Flood.
17 See Isa. 66:22; 2 Pet. 3:13. The prophet Moroni reasoned that Jerusalem in Palestine could not be the New Jerusalem "for it had been in a time of old; but it should be built up again" (Ether 13:5). So will it be with Earth in the millennial period; like old Jerusalem it will be the same core planet, but gloriously renewed and built up again.
18 See D&C 84:33; JD, 15:365.

19 See JD, 1:331. The very fact that Earth has to be born again into the Lord's presence is good evidence that it was in his presence when it was first organized. As Son of Man, Jesus Christ is the God who ministers to the terrestrial kingdom (see D&C 76:71, 77). The millennial planet was seen in vision by the ancient apostles, Peter, James and John. They were shown "the pattern" by which the transfigured Earth would receive its paradisiacal glory at the time they beheld the transfiguration of Jesus (see D&C 63:20-21;Mark 9:2-3; 2 Pet. 1:16-18).
20 See Isa. 30:26. The number seven denotes "many" and is not always interpreted literally.
21 See Dan. 3:19-25.
22 See D&C 86:10; Acts 3:21.
23 See D&C 1:1, 38; 39:16.
24 See JST Gen. 14:34-35.
25 See 9:2; Rev. 6:14; 16:20.
26 N.L. Lundwall, *Inspired Prophetic Warnings,* 38.
27 See Isa. 24:1, 19-20.
28 That is, "My delight is in her."
29 That is, "Married."
30 Joseph Smith quoted with approval the first issue of *The Evening and Morning Star* concerning the land of Zion being "joined, or married. . . to Jerusalem again, and they be one as they were in the days of Peleg" (DHC, 1:275).
31 There is no agreement among geologists as to the actual configuration of Pangaea.
32 "Island" or "isles of the sea," may refer to entire continents as well as lesser land masses. For example, the Nephites in America spoke of themselves as being "upon an isle of the sea" (2 Ne. 10:20). Satan's "mother of abominations" sits upon "the islands of the sea" (D&C 88:94; Rev. 17:1-5).
33 The present Atlantic Ocean will virtually disappear when North and South America rejoin Europe and Africa—the relationships which existed originally according to substantial geologic evidence.
34 See Gen. 7:11; 8:2.
35 See Abr. 4:6-8; see Moses 2:6-8.
36 See Isa. 65:20; D&C 63:51; 101:30.
37 See Luke 3:5; D&C 49:23.
38 See Isa. 54:10; 64:1-3; Micah 1:4; Rev. 6:14; 16:20; 3 Ne. 22:10; D&C 109:74; 133:22, 40, 44: TPJS, 12-13.
39 Journal of Samuel H. Rogers, Brigham Young University, H. B. Lee Library, 17.
40 Eliza R Snow reiterated these ideas in private conversations over a period of about thirty years. Charles L. Walker quoted her as saying the

earth was ninety times smaller than when first organized (Walker, 2:540).

41 See Mal. 4:2; 2 Chron. 7:14. Orson Pratt noted: "When it rains upon the exalted valleys, it will wash down the rich soil upon the rocky mountains which have sunk beneath, making them fertile" (JD, 18:320).

42 See JD, 1:203; 10:301.

43 Enoch's Zion, together with the Zion of Joseph in America and of Judah in Jerusalem, constitute, in their union, the nucleus of the millennial society which the Savior will establish on a world-wide basis.

44 See Psalm 67; Ezekiel 34:27 and TPJS, pp.248-49.

45 See D&C 49:19.

46 See Dan. 2:31-45.

47 See James 4:1-2.

48 See Eph. 2:14.

49 See D&C 78:5-6.

50 See 2 Ne. 30:10-15.

51 Orson Pratt suggested: "And then the animal creation will manifest more intelligence and more knowledge than they do now, in their fallen condition. . . .they will even know how to praise God . . . .What? The animal creation endowed with language? Yes, a language of praise" (JD, 20:18; see Rev. 5:13-14).

52 See Isa. 2:4; Ezek. 34:25; Hosea 2:18; JD, 19:175; 20:18, 21:203; 24:210.

53 *Times and Seasons,* 3 [1 June 1842], 799-800); see JD, 20:18.

54 Whether longevity in the millennial period will equal that of the pre-Flood patriarchs who lived for centuries is doubtful. But there will be no infant mortality; all will live to a ripe old age. Bruce R. McConkie anticipated a pre-Fall, Edenic, condition for mankind, writing that "man and beast are changed (quickened) and blood no longer flows in their veins" (*The Millennial Messiah,* 658; see 654).

55 See D&C 63:51; Isa. 65:20, 22. At last there will be one less meeting to attend!

56 See JD, 18:290.

57 The paragraph from which this quote was taken first appeared in an editorial written by John Taylor for the *Times and Seasons* (15 Feb. 1845)—a response to W.W. Phelps' earlier editorial (1 Jan. 1845) in that newspaper.

58 See D&C 88:51-61; 29:30-33.

59 Orson Pratt identified the twelve kingdoms with the other planets of our solar system which were also hidden from the Lord's presence. Hence, the need to visit them (see JD, 19:293-294). In 1911 Benjamin F. Johnson, a confidant of Joseph Smith, wrote of this concept to George Gibbs: "He [Joseph Smith] gave us to understand that there were

twelve kingdoms, or planets, revolving around our solar system, to which the Lord gave equal division of His time or ministry; and now was his time to again visit the earth" (Letter of Benjamin F. Johnson to George S. Gibbs, compiled by Chas. S. Sellers, 1 July 1911, Mesa, Arizona).

60 See D&C. 90:3.
61 See D&C 38:17; 88:19.

CHAPTER TWELVE

THE CELESTIAL EARTH

The Little Season

As the sun sets on Earth's second sabbath, the angel of the Lord will turn the key, Satan will be released from his long imprisonment, and spiritual darkness will descend again upon the world.[1] The "little season" has begun.[2] Those living at that time will face Lucifer's opposition in his last desperate, but futile, effort to overthrow his brother, Jesus Christ. In spite of possessing unsurpassed spiritual knowledge, many will succumb and wilfully rebel against the truth in the most widespread apostasy since the war in heaven. There will be many fatalities.[3] Satan will deceive the nations as he deceived them a thousand years earlier. They will marshall their forces (Gog and Magog) and again lay siege to "the beloved city."[4] This will be the climactic struggle between the forces of good and evil on this planet:

> And Michael, the seventh angel, even the archangel, shall gather together his armies, even the hosts of heaven. And the devil shall gather together his armies; even the hosts of hell, and shall come up to battle against Michael and his armies. And then cometh the *battle of the great God;* and the devil and his armies shall be cast away into their own place, that they shall not have power over the saints any more at all. For Michael shall fight their battles, and shall overcome him who seeketh the throne of him who sitteth upon the throne, even the Lamb (D&C 88:112-115; see Rev. 20:7-10).

Thus will Michael who, as Adam, faced and defeated Satan in the beginning of time, face and defeat him in the end of time.[5] Lucifer and his hosts will thereafter be banished to perdition, the second death. The throne of the Son of Man will never be threatened again. With the close of the little season, the time will have come for Earth to pass away.[6]

The Last Resurrection

> But, behold, verily I say unto you, *before* the earth shall pass away, Michael, mine archangel shall sound his trump, and then shall all the dead awake, for their graves shall be opened, and they shall come forth—yea, *even all* (D&C, 29:26; see. 88:101).

The principle of the last being first also applies to Earth. Her offspring, both man and creature, will be resurrected before she is. Hence, before this planet is reduced to disorganized element, every order of life destined for immortality will be removed from it. Were this not the case, those essential elements belonging to the body would become lost or incorporated into the planet itself when it is immortalized. This would frustrate the Redeemer's plan for a perfect resurrection of all things.

But the *material* integrity of the human body is inviolate. Its *essential* identity is never lost; its unique components never become identified permanently with any other organism. Joseph Smith taught:

> There is *no fundamental principle* belonging to the human system that ever goes into another in this world or in the world to come.... If anyone supposes that any part of our bodies, that is, the fundamental parts thereof, ever goes into another body, he is mistaken (DHC, 5:339).[7,8]

Through the power of Christ, spirit and body are joined forever:

> The soul [spirit] shall be restored to the body, and the body to the soul; yea, and every limb and joint shall be restored to its body; yea, even a hair of the head shall not be lost; but all things shall be restored to their proper and perfect frame (Alma 40:23; see D&C 29:24-25).

This is done by one exercising Priesthood keys of resurrection. Brigham Young:

> Some person holding the keys of the resurrection, having previously passed through that ordeal will be delegated to resurrect out bodies, and our spirits will be there prepared to enter into their bodies (JD, 9:139; see D&C 88:101).

The Death of Earth

Following the last resurrection, this planet will become, at its death, as devoid of life as it was at its birth. So far as the sum of mankind is concerned, Earth's mission will be over; it will have filled the temporal measure of its creation. How will this planet die? The manner of its passing is a clue to the time of its passing. Formerly astrophysicists theorized it would be either destroyed when the Sun exploded or else become a frozen hulk when the Sun's energy supply was exhausted, causing the sun to sputter and die like a candle flame. Current theory maintains that the Sun, a "middle-aged star," will increase in luminosity (and hence temperature) until it becomes a "red giant."[9] It is then expected to radiate enough energy to virtually incinerate the earth. This is known as the "heat death;" it is scheduled to occur in several billion years.[10] However, this planet's passing is programmed a bit sooner than that; it will survive only another thousand years or so.

John the Revelator testified: "And I saw a great white throne, and him that sat on it, from whose face the *earth and the heaven fled away;* and there was found no place for them" (Rev. 20:11).[11]

The demise of this planet is spoken of in modern revelation as "the end of the earth."[12] The most comprehensive statement in all scripture is the following:

> And the end shall come, and the heaven and the earth shall be consumed and pass away, and there shall be a new heaven and a new earth. For all old things shall pass away, and all things shall become new, even the heaven and the earth, and all the fulness thereof, both men and beasts, the fowls of the air, and the fishes of the sea (D&C 29:23; see Ether 13:9; Rev. 21:1).[13]

Earth will cease to exist as an organized planet, having been vaporized "so as by fire."[14] But it will not be cremated by blind circumstance in some future fire storm on the sun. Its passing will be far more than the bursting of an insignificant bubble in the cosmic cauldron of chance events. Christ, not chance, will bring about its death; it will vanish away "by the power of his almighty word" (1 Ne.17:46). All that remains will be the chaotic matter from whence it came. Earth will have returned to native element.

Resurrection of Earth

Mortal men and worlds are borrowers of light, dependent for illumination upon things, both natural and artificial, outside of themselves. As long as the spirit remains separate and distinct from the body, the body will remain a borrower of light. For spirit is light, the fountain of glory. Only when the spirit is bonded to the body, permeating every cell thereof, can the body become light-filled, spirit-filled and, therefore, glorious in and of itself. This brings to pass radiant worlds and radiant men and women—the divine transformation from mortal darkness to immortal luminescence.[15]

Therefore, after Earth and its heaven have "fled away," it will experience its ultimate and most sublime rebirth—resurrection. As with its creation, its resurrection will not be a matter of billions of years; virtually in the twinkling of an eye Earth will be reorganized into an *immortal star!* According to known physical laws, such a thing is an utter impossibility. Nevertheless, at the command of its Creator it will rise from its grave complete and whole, its *every particle* restored—spirit and element being inseparably connected forever. All this through the transcendent governing laws of the Almighty. It will be endowed with a fulness of celestial glory, a glory which can only be endured by those capacitated to dwell in its "everlasting burnings":

> That bodies who are of the celestial kingdom may possess it forever and ever; for, for this intent was it made and created, and for this intent are they sanctified (D&C 88:20).

Earth is a covenant planet designed from the beginning to become a great celestial kingdom.[16] It was never destined for a lesser glory; therefore, it was never placed under the law of a lesser order of worlds.[17] Its fall to a telestial condition drove it from God's presence and it became a temporal sphere with all of its present imperfections and limitations:

> Wherefore, it shall be sanctified; yea, notwithstanding it shall die, it shall be quickened again, and shall abide the power by which it is quickened, and the righteous shall inherit it (D&C 88:25-26).

No longer will it beg the borrowed light of lesser worlds. No longer will it be the slave of the Sun, held captive by its gravitational powers and dependent upon its life-sustaining rays. Rotating in harmony with Kolob and its exalted companions, its days will be once more a thousand years in length.[18] It will be an immortal Star in its own right. Brigham Young remarked:

> By and by the Lord will purify the earth, and it will become pure and holy, like a sea of glass; then it will take its place in the rank of the celestial ones, and be recognized as celestial; but at the present time it is a dark, little speck in space (JD, 14:136; see 17:117).

Having fled this benighted solar system, Earth will join its celestial peers:

> And thou, O Earth, wilt leave the track
>   Thou hast been doom'd to trace—
> The Gods with shouts will bring thee back
>   To fill thy native place.[19]

The Presence of the Father

In returning to its "native place," the Celestial Earth passes through the veil and enters the spiritual presence of the Father:

> Therefore, it must needs be sanctified from all unrighteousness, that it may be prepared for the celestial glory; For after it hath filled the measure of its creation, it shall be crowned with glory, even with the presence of God the Father (D&C 88:18-19).

"This earth," said the Prophet Joseph, "will be *rolled back* into the presence of God, and crowned with celestial glory" (TPJS, 181; emphasis added).[20] Brigham Young agreed, it "will be brought into the immediate presence of the Father and the Son" (JD, 8:200). He later remarked, "when it is celestialized it will go back into the presence of God, *where it was first framed*" (JD 9:317).[21]

Though the immortal Earth may be trillions of miles distant from his literal residence, nevertheless it will be in the Father's living presence. Modern communications provide the merest suggestion of how direct and intimate will be the association of the Father with his children though they should dwell in different eternal galaxies in the universe. Orson Pratt anticipated just such a condition:

> I have no doubt that many of us will be counted worthy to approach near to him so far as distance is concerned. But then, when we come to reflect that distance will be comparatively annihilated, between God and the worlds he has made, so that it will make no difference, as far as his presence is concerned, whether he is close by or millions of miles distant—there will be a mutual communication between the Creator and his children all the time, consequently there will be union and fellowship with him, and rejoicing in his presence, though he be in a world far beyond Kolob, of which Abraham speaks (JD, 15:239).

A Sea of Glass

John the Revelator beheld God sitting upon his throne with a "sea of glass like unto crystal" lying before him (Rev. 4:6).[22] Eighteen centuries later Joseph Smith learned the sea of glass represented "the earth, in its sanctified, immortal and eternal state" (D&C 77:1).[23] "a globe like a sea of glass and fire, where all things are manifest, past, present, and future, and are continually before the Lord" (D&C 130:7-8).

As mortals living on a planet with varied seasons, climatic conditions, landscapes and all else that is familiar to us, a description of heaven as a "sea of glass" and "eternal burnings" hardly sounds appealing, but we must realize that we are totally incapable of comprehending heaven as it is, therefore the Lord describes it with analogies: "God dwells on a globe *like* a sea of glass;" the gate to the celestial kingdom "was *like* unto circling flames of fire."

These descriptive terms are colored by our own situation. The sea, so cold and inhospitable to man, is the natural environment of marine life. So, too, heaven is glass and fire from our perspective, not from that of those dwelling there. As immortals, we will be far more comfortable in heaven's "eternal burnings" than we were on our most pleasant day on Earth. The "burnings" of the Celestial Kingdom merely reflect the glory-filled natures of its inhabitants.

Although bathed in immortal glory, the celestial world will not be a strange, unfamiliar creation; like its resurrected inhabitants, it "shall be *like unto the old* save the old have passed away, and all things have become new" (Ether 13:9). It will be a realm of dazzling natural beauty where exquisite trees, flowers, and shrubbery of seemingly endless variety abound. Magnify all the lovely things of this planet a thousand times over and we would still fail to comprehend the splendor of the eternal Earth.[24] Its natural beauty will be accented by the presence of resurrected creatures of every kind. They will fly through its skies, swim in its waters, and walk upon its breast again.

Beholding the Celestial Earth, John wrote, "there was *no more sea*" (Rev. 21:1). The planet's present oceans will not be found on the resurrected Earth; they could not exist in a realm of "eternal burnings." However, there will be celestial waters. John's "fellow servant," the prophet-angel of his revelation, showed him "a pure river of water of life, clear as crystal, proceeding out of the throne of God and of the Lamb" (Rev. 22:1-2). Since marine life will also be resurrected, celestial waters will be provided for at least some of it. The Lord loves variety, and it abounds throughout all of his creations. Joseph Smith commented:

John saw. . . .every creature that was in heaven,—all the beasts, fowls and fish in heaven,—actually there, giving glory to God. . . .The grand secret was to show John what there was in heaven. John learned that God glorified Himself by saving all that His hands had made, whether beasts, fowls, fishes or men; and He will glorify Himself with them. Says one, "I cannot believe in the salvation of beasts." Any man who would tell you that this could not be, would tell you that the revelations are not true (TPJS, 291).[25]

Powers of Celestial Beings

The resurrection will not only restore, but vastly expand the intellectual faculties of those who dwell on the celestial Earth.[26] For it will constitute a mighty Urim and Thummim, a revelator of all things pertaining to *lesser* worlds:

> This earth, in its sanctified and immortal state, will be made like unto crystal and will be a Urim and Thummim to the inhabitants who dwell thereon, whereby all things pertaining to an inferior kingdom, or all kingdoms of a lower order, will be manifest to those who dwell on it; and this earth will be Christ's (D&C 130:9; see JD, 14:136).[27]

The Prophet Joseph taught that when Earth was celestialized, "the Saints could look in it and see as they are seen" (DHC, 5:279).[28] Brigham Young expanded on this doctrine, stating "when you wish to know anything, you can look in this earth and see all the eternities of God." (JD, 8:200). He described the experience as rather like looking in a mirror:

> You step up and you look and you see Adam as he was. You would like to see Jerusalem as it looked when the temple was finished, I walk up to this looking glass, and I look. Well [when] the whole earth becomes like that it will be a true mirror, or Urim and Thummim. And everything you

desire to see or know is present before you; you look and it is there.[29]

Two Celestial Cities

When Earth is endowed with celestial glory, the two holy cities from which the law of the Lord was sent forth to the nations in the millennium will return to their ordained places. The Revelator wrote that he saw the "new Jerusalem, coming down from God out of heaven prepared as a bride adorned for her husband" (Rev. 21:2). Thereafter, an angel showed him a second city: "the holy Jerusalem, descending out of heaven from God" (Rev. 21:10; see JD, 18:346-48). Ether, the last of the Jaredite prophets, foretold these cities:

> The earth shall pass away. And there shall be a new heaven and a new earth. . . .And then cometh the New Jerusalem; and blessed are they who dwell therein, for it is they whose garments are white through the blood of the Lamb...And then *also* cometh the Jerusalem of old: and the inhabitants thereof, blessed are they, for they have been washed in the blood of the Lamb. . . .And when these things come, bringeth to pass the scripture which saith, there are they who were first, who shall be last; and there are they who were last, who shall be first (Ether 13:10-12).[30]

The New Jerusalem, having been established last, will be given precedence over the Jerusalem of old—the last shall become first in descending to the immortal Earth. Beholding the holy Jerusalem, the apostle John described it as being of "pure gold, like unto clear glass" (Rev. 21:18, 21).[31] "The city had no need of the sun, neither of the moon, to shine in it: for the glory of God did lighten it, and the Lamb is the light thereof" (Rev. 21:23).[32] Neither Earth nor its inhabitants will be obliged to borrow light any longer, for they will be filled with the glory of God—they will be as those worlds which never fade and never die, realms of pure spiritual light.

The God of Earth

Lucifer is the self-appointed "god of this *world*," but Jesus Christ is the God of this *Earth*.[33] He created it, sustains it, redeems it, and will resurrect it. When it achieves its exalted celestial status he will rule it— "this earth will be Christ's" (D&C 130:9). He will become a "Father" in his own right, possessing and ruling over this celestial earth as part of the inheritance bestowed upon him by his Father in this eternity. This truth was revealed by the Prophet Joseph Smith in his most profound extant sermon, the "King Follett Discourse" of 7 April 1844:

> What did Jesus do? Why; I do the things I saw my Father do when worlds came rolling into existence. My Father worked out his kingdom with fear and trembling, and I must do the same; and *when I get my kingdom,* I shall present it to my Father, so that he may obtain kingdom upon kingdom, and it will exalt him in glory. He will then take a higher exaltation, and *I will take his place*, and thereby become exalted myself.[34] So that Jesus treads in the tracks of his Father, and *inherits what God did before;* and God is thus glorified and exalted in the salvation and exaltation of *all* of his children (TPJS, 347-48; emphasis added).

Herein is the very heart of the Gospel of Jesus Christ, of the law of Consecration and Stewardship, and of the Patriarchal Order of the Holy Melchizedek Priesthood. Each serving each with an eye single to the glory of the Father who, in ascending ever higher, draws his faithful sons and daughters upward to ever greater and greater dominion and glory! The Son of God, having consecrated all to his Father and his God, receives in return not merely a stewardship, but *an inheritance—Earth!*

The love of God is astonishing! Jesus Christ not only suffered the incomprehensible pains of the Atonement to save us from our sins, he suffered so that he could share his inheritance, as the Father's sole heir, with us![35] We become joint-heirs with Christ by his grace. Such is the divine law of consecration.

The Meek Shall Inherit the Earth

"Blessed are the meek: for they shall inherit the earth" (Matt. 5:5). Such was the promise of Jesus to his disciples some nineteen centuries ago. Cynical men continue to ridicule these words by pointing out that it is not the meek, but the mighty, who have dominated the earth since Cain murdered Abel. But they err, not knowing the scriptures nor the mind of the Lord. Jesus well understood the character of this world with its fawning obeisance to power, whatever its guise. He knew that might will continue to make for right as long as men and devils dictate the affairs of mankind. Indeed, in his beatitude, Jesus was quoting an ancient psalmist who had written those words with the very problem of injustice in mind:

> But the meek shall inherit the earth; and shall delight themselves in the abundance of peace. . . . .I have seen the wicked in great power, and spreading himself like a green bay tree. Yet he passed away, and, lo, he was not: yea, I sought him, but he could not be found (Ps. 37:11, 35-36).

Neither Jesus nor the psalmist was alluding to the meek or the Earth in their fallen states. The promise does not pertain to either of them as they are, but only as they shall be. Death precedes inheritance. This is doubly true in this instance: the death of the meek and the death of the Earth. Both must be made glorious before the inheritance can be claimed. President John Taylor explained, "if you possess any portion of this earth by right or title or authority, you will have to get it from God, and you will have to get it *when the earth shall be renewed"* (JD, 22:220).[36]

The immortal Earth is the true and everlasting land of promise of the saints of God of all ages. Lands which have been designated as "promised lands" are but symbols of what is to be. These promised lands can only be received by covenant, otherwise there is no inheritance. Only those who are joint-heirs with Jesus Christ in all that the Father possesses, have a legal claim to Earth after they are dead. Orson Pratt spoke to this point:

> In the resurrection, the meek of all ages and nations will be restored to that portion of the earth previously promised to them. And thus, all the different portions of the earth have been and will be disposed of to the lawful heirs; while those who cannot prove their heirship to be legal, or who cannot prove that they have received any portion of the earth by promise, will be cast out into some other kingdom or world, where, if they ever get an inheritance, they will have to earn it by keeping the law of meekness during another probation. How great will be the disappointment to the rich, the high and the noble, who have rejected the messages of eternal truth, sent forth in different ages for the redemption of men, when they find that there is not a foot of the new earth that they can call their own; the whole of it having been lawfully disposed of to the poor and the meek (JD, 1:332-33).

Brigham Young emphasized that the meek shall not enter into their inheritances until Jesus Christ is crowned King of kings—King over all those for whom he made exaltation possible:

> [When] the Savior has presented the earth to his Father, and it is placed in the cluster of the celestial kingdoms, and the Son and all his faithful brethren and sisters have received the welcome plaudit— 'Enter ye into the joy of your Lord,' and the Savior is crowned, then and not till then, will the

Saints receive their everlasting inheritances (JD, 17:117; see 6:282).[37]

The risen Redeemer told the surviving remnant of the Nephite nation, "I would that ye should be perfect even as I, or your Father who is in heaven is perfect" (3 Ne. 12:48). Perfection is relative to one's state; it is attainable in every future heaven or degree of glory.[38] But only those who achieve celestial perfection are heirs of a celestial salvation. And of that number, only those men and women who attain celestial perfection in its most exalted form will be *capacitated* to rule as kings and queens over their own endless posterity. This state of perfection requires more than repentance and a remission of sins, more than routine, mechanical conformity to gospel principles. It demands *becoming*. Only those who become like God will attain it.

The Prophet Joseph Smith taught:

> . . .you have got to *learn how to be Gods yourselves*. . . . [being] *able* to dwell in everlasting burnings, and to sit in glory, as do those who sit enthroned in everlasting power (TPJS, 346-47; emphasis added).

He rendered John 14:2: "In my Father's kingdom are many kingdoms," and went on to say:

> There are mansions for those who obey a celestial law, and there are other mansions for those who come short of the law every man in his own order (TPJS, 366).

Those who are "*not able* to abide the law of a celestial kingdom cannot abide a celestial glory" (D&C 88:22).[39] Consequently, such men and women will be obliged to bid Mother Earth an everlasting farewell; their mansions lie elsewhere in the Father's vast dominions.

But the meek—perfected in light and truth, *becoming capacitated* to dwell in celestial glory—shall inherit the immortal Earth. And Jesus Christ, the Creator, having organized, redeemed, and glorified this planet, will claim it for his own. Earth—presently the footstool of the Son of Man—will have attained the ultimate measure of its creation and become his very throne!

Epilogue

And so another cycle of creation is completed, another Earth redeemed, another family of Man perfected, another victory for righteousness and truth.

> And I heard a great voice out of heaven saying, Behold, the tabernacle of God is with men, and he will dwell with them, and they shall be his people, and God himself shall be with them, and be their God. And God shall wipe away all tears from their eyes; and there shall be no more death, neither sorrow, nor crying, neither shall there be any more pain: for the former things are passed away (Rev. 21:3-4).

Happiness— "the object and design of our existence"[40]—will prevail throughout the kingdom of God in all of its heavens. But its highest expression will be found among those sons and daughters of the Father who realize a celestial exaltation. Endowed with eternal lives, possessing the procreative powers of their divine Parents, they will be capacitated to join them in bringing to pass the immortality and eternal life of man. For them, the command to multiply and replenish the earth will never be revoked. Their kingdom is their posterity; it will be added upon forever and ever. They are Gods.[41]

In the aeons to come they will organize new Earths from eternity's infinite materials.[42] The spirit offspring—organized intelligences—of these heavenly Fathers and Mothers will follow in their steps and make the long journey to another fallen world where they too will experience the essential realities of a mortal existence. And when they go, they will take something of this celestial realm with them. For even as Mother Earth in her fallen state provided their once-mortal parents with physical bodies, so will she, in her glory, provide these intelligences with spirit bodies. In this way, an endless chain of life is formed, link by link, binding the family of Man together from world to world, and from eternity to eternity.

The "eternal round" of the Gods will begin again. The curtain will rise on yet another production of the endless cosmic drama. New players will perform old, familiar parts. Ancient actions will be repeated as ancient dialogue falls again from the lips of saints and sinners, wise men and fools. And when the final curtain comes down, truth, justice and mercy will have triumphed once more over human frailty and the powers of evil. Another perfected, immortal generation of the family of Man will be a reality. And another realm of immortal glory will find its place in the galaxy of celestial worlds to which *This Eternal Earth* will have long since ascended.

NOTES

1 See Rev. 20:1-3, 7-8; D&C 29:22; 88:110-111; TPJS, 65.
2 The term, "little season" implies a brief period of time. But it is imprecise since it is associated with the martyrs awaiting resurrection (Rev. 6:11), the redemption of Zion (D&C 100:13; 103:44; 105:9, 13), groups and individuals (D&C 42:5; 51:16;.63:42; 88:71; 105:21), and the postmillennial period (Rev. 20:3; D&C 29:27; 43:31; 88:111). As to the latter, some believe it will last a thousand years, a very doubtful possibility. It may approximate the period of time Adam and Eve were in the Garden of Eden prior to the Fall.
3 The number of spiritual fatalities—sons of perdition—in the "little season" may exceed the number resulting from the war in the first estate.
4 See JD, 16:322-23.
5 Adam did not succomb to Satan's temptation; he deliberately partook of the forbidden fruit "that men might be" (see 1 Tim. 2:14; 2 Ne. 2:25).
6 See D&C 43:31.
7 See *WJS*, 182.
8 Brigham Young explained, "the elements will be burned and purified and be renewed; but not one atom of earth's organism will be lost; for that which is governed by law shall be preserved by law" (Conference address, 8 October 1875, Church Archives).
9 It is believed that following the "red giant" stage, the sun will become a "white dwarf" and, after exhausting its nuclear energy supply, become a dark burned-out voyager through space. It will have "lived" for about ten billion years.
10 All life on Earth would have been destroyed long before this occurred.
11 See JD, 16:322. The "white throne judgment," is the last judgment. It takes place following the end of the earth when the entire resurrected human race returns to the literal presence of God (see D&C 38:5; 2 Ne. 9:38; Hel. 14:17; Alma 42:23).
12 "End of the earth" is a phrase unique to the revelations of Joseph Smith (see D&C 38:5; 43:31; 45:26; 88:101; 107:42; JST Matt. 24:56).
13 The "end of the earth" should not be confused with the end of the world which refers to the destruction of the telestial order at Christ's coming (see JST, Matt. 24:4, 56; D&C 45:22).
14 D&C 43:32. Orson Pratt asked: "What is it that will make the earth die? It will be the withdrawing of the spiritual portion from it, that which gives it life—that which animates it, and causes it to bring forth fruit; that which quickens the earth is the Spirit of God" (JD, 21:201; see. JD 21: 2:340).

15 However, the magnitude of one's immortal radiance, or degree of glory, depends upon the amount of intelligence, the light and truth, acquired by one's spirit during the probationary period prior to resurrection.
16 See D&C 88:25.
17 See D&C 88:36-39.
18 Earth may begin its return journey toward Kolob when it is renewed with paradisiacal glory. Isaiah's reference to the sun's light becoming sevenfold brighter in the millennium is noteworthy in this regard (see Isa. 30:26).
19 Snow, "Address to Earth." Br [Addison] Everett said that he heard Joseph say that "the earth had been divided and parts taken away, but the time would Come when all would be restored and the earth would revolve in its original orbit next to Kolob and would be second in size to it" (Charles Walker Journal, [18 Oct. 1880] 2:505).
20 See JD, 8:8, 200; 9:292.
21 See 7:163; 17:143.
22 See 15:2; 21:18, 21.
23 See JD, 8:200; 14:136; 16:323. Silica, soda, lime, and potash are so plentiful that if the earth were melted down "you would end up with a ball of glass" (*Science Digest Special,* (July 1967), 45.
24 Brigham Young described the present world as "nothing but a garden spot in comparison to the rest of the kingdoms of our God" (JD, 4:268).
25 See Rev. 4:6-9; 5:6-14; D&C 77:2-4. The ultimate progress of any organized intelligence lies in its willingness and ability to submit to and unite with a greater intelligence (see D&C 88:40). In this respect, the Earth and many of her creatures, will be blessed in their "eternal felicity" above many a child of God. Is it not ironic that certain classes of animals will be found in the celestial kingdom while certain classes of men will not?
26 On a number of occasions Orson Pratt discussed the future powers of man (see JD, 3:97-105; 19:177-178, 292-294; 21:256-263, 328).
27 A personal Urim and Thummim will be given to the exalted inhabitants of the celestialized Earth "whereby things pertaining to a higher order of kingdoms will be made known" (D&C 130:10).
28 See 1 Cor. 13:12.
29 Discourse, 17 Mar. 1861, Church Archives.
30 In commenting on Ether's prophecy. Joseph Smith wrote: "Now we learn from the Book of Mormon the very identical continent and spot of land upon which the New Jerusalem is to stand, and it must be caught up according to the vision of John upon the isle of Patmos. Now many will feel disposed to say, that this New Jerusalem spoken of, is the Jerusalem that was built by the Jews on the eastern continent. But you will see, from Revelation xxi: 2, there was a New Jerusalem

coming down from God out of heaven, adorned as a bride for her husband; that after this the Revelator was caught away in the Spirit, to a great and high mountain, and saw the great and holy city descending out of heaven from God. Now there are two cities spoken of here" (TPJS, 85- 86; emphasis added; see Rev. 3:12; JD, 1:332; 16:323; 20:15).

31 Joseph Smith described the streets of the celestial kingdom as having the "appearance of being paved with gold" (D&C 137:4).

32 See Ether 13:3, 10. Orson Hyde assumed the sun would continue to give light to the earth (see JD, 1:130).

33 See 2 Cor. 4:4.

35 See Heb, 1:2; Rom. 8:14-17.

36 See 1:293; 14:237-238; 20:16-17.

37 According to Brigham Young, Joseph Smith told John Taylor that his, Joseph's, inheritance would be located near Nauvoo, Illinois (Discourse, 17 March 1861, Church Archives).

38 Brigham Young observed, "When we use the term perfection, it applies to man in his present condition, as well as to heavenly beings. We are now, or may be, as perfect in *our sphere* as God and Angels are in theirs, but the greatest intelligence in existence can continually ascend to greater heights of perfection" (JD, 1:93; emphasis in original).

39 Brigham Young observed, "There is a great difference in the individual capacity of people. Some can receive much more than others can; hence we read of different degrees of glory" (JD, 9:106-07).

40 TPJS, 255.

41 Joseph Smith declared, "every man who reigns in celestial glory is a God to his dominions" (TPJS, 374).

42 Brigham Young cautioned; "We can have no other kingdom until we are prepared to inhabit this [earth] eternally" (JD, 3:372).

BIBLIOGRAPHY

Andrus, Hyrum L., *God, Man, and the Universe* (Salt Lake City: Bookcraft, 1968)
—and Helen Mae, *They Knew the Prophet* (Salt Lake City: Bookcraft, 1974)
Benson, Ezra Taft, *Teachings of Ezra Taft Benson* (Salt Lake City: Bookcraft, 1988)
Clark, James R., *Messages of the First Presidency* (Salt Lake City: Bookcraft, 1965)
Clark, J. Reuben, *Behold the Lamb of God* (Salt Lake City: Deseret Book, 1991)
Cowley, Matthias, *Wilford Woodruff* (Salt Lake City: Bookcraft, 1964)
Dalton, M.W., *A Key To This Earth* (Willard, Utah: 1906)
Hinckley, Bryant S., *Sermons and Missionary Experiences of Melvin J. Ballard* (Salt Lake City: Deseret Book, 1949)
Ehat, Andrew F. and Cook, Lyndon W., *The Words of Joseph Smith* (Provo: Religious Studies Center, 1980)
Johnson, Benjamin F., *My Life's Review* (Independence: Zion's Publishing, 1947)
Journal of Discourses, 26 vols. (London: Latter-day Saints' Book Depot, 1856-86)
Larson, A.K. and K.M. Larson, eds., *Diary of Charles L. Walker* (Logan: Utah, State University Press, 1980).
Lundwall, N.B., *The Vision* (Independence: Zion Publishing, 1945)
—*Inspired Prophetic Warnings* (Salt Lake City, 1940)
Mace, Wandle, *Autobiogrphy of Wandle Mace,* manuscript copy, Special Collections, Harold B. Lee Library, Brigham Young University
Maxwell, Neil A., *But For A Small Moment* (Salt Lake City: Bookcraft,1986)
Matthews, Robert J., *Joseph Smith's Translation Of The Bible* (Provo: Brigham Young U. Press, 1985)
McConkie, Bruce R., *Mormon Doctrine* (Salt Lake City: Bookcraft,1966)
—*The Millennial Messiah,* (Salt Lake City: Bookcraft, 1982)
—*A New Witness for the Articles of Faith* (Salt Lake City: Deseret Book, 1985)
Nibley, Hugh, *Old Testament And Related Studies* (Salt Lake City: Deseret Book, 1986)
Pratt, Orson, *The Seer* (London: F.D. Richards, 1853)
Pratt, Parley P., *A Voice of Warning*
—*Key to Science of Theology*

Reynolds, George, *Are We of Israel?* (Independence, Zion's Publishing, n.d.)
Richards, Franklin D., *A Compendium* (Salt Lake City: Cannon & Sons, 1898)
Roberts, B.H., *The Gospel and Man's Relationship to Deity* (Salt Lake City: Deseret Book, 1946)
—*The Truth, the Way, The Life* (Provo: BYU Studies, 1994)
—*Comprehensive History of the Church* (Salt Lake City: LDS Church, 1930)
Rogers, Samuel Holister, *Diary of Samuel Holister Rogers,* typescript, Special Collections, Harold B. Lee Library, Brigham Young University
Smith, George D., ed., *The Journal of William Clayton,* (Salt Lake City: Signature Books, 1995)
Smith, Joseph, *Lectures on Faith* (Salt Lake City: Deseret Book, 1985)
Smith, Joseph F., *Gospel Doctrine* (Salt Lake City: Deseret Book, 1946)
Smith, Joseph Fielding, *Doctrines of Salvation* (Salt Lake City: Bookcraft, 1954-56)
—*Man, His Origin and Destiny* (Salt Lake City: Deseret Book, 1954)
—*Signs of the Times* (Independence: Zion's Publishing, 1947)
—*Teachings of the Prophet Joseph Smith* (Salt Lake City: Deseret Book, 1938)
—*The Way to Perfection*, 1947 (Independence: Genealogical Society, 1946)
Smith, Robert W., *The Last Days* (Salt Lake City: Author, n.d.)
Smith, Hyrum M. and Sjodahl, Janne M., *Doctrine and Covenants Commentary* (Salt Lake City: Deseret Book, 1950)
Snow, Eliza R., *Poems, Religious, Historical, and Politcal* (Liverpool:1856)
Stuy, Brian H., *Collected Discourses* (Sandy, Utah: B.H.S. Publishing, 1987)
Talmage, James E., *The Articles of Faith* (Salt Lake City: LDS Church, 1946)
Taylor, John, *The Gospel Kingdom* (Salt Lake City: Bookcraft, 1944).
—*Mediation and Atonement*, (Salt Lake City: Stevens & Wallis, 1950)
The Latter Day Saints Millennial Star (Liverpool: 1840-1970)
Times And Seasons (Nauvoo: 1839-46)
Turner, Rodney, *Woman and the Priesthood* (Salt Lake City: Deseret Book, 1972)
Whitney, Orson F., *Saturday Night Thoughts* (Salt Lake City: Deseret Book, 1921)
Widtsoe, John A., *Discourses of Brigham Young* (Salt Lake City: Deseret Book. 1946)

Woodruff, Wilford, *Wilford Woodruff's Journal,* 1833-1898, typescript, (Midvale, Utah: Signature Books, 1985)
—*Gospel Interpretations* (Salt Lake City: Bookcraft, 1947)

TECHNICAL PUBLICATIONS

Cook, Melvin A., *Scientific Prehistory* (Salt Lake City: Family History, 1993)
—and Melvin G., *Science and Mormonism* (Salt Lake City: Deseret Book, 1973)
Ferris, Timothy, *Coming of Age in the Milky Way* (New York: William Morrow, 1988)
Henry Frankfort, *The Birth of Civilization in the Near East* (Bloomington: Indiana University Press, 1951)
Jastrow, Robert, *Red Giants and White Dwarfs* (The New American Library, 1969)
Jastrow, Robert and Thompson, Malcolm H., *Astronomy: Fundamentals and Frontiers* (New York: John Wiley & Sons, 2nd. ed., 1974)
Morris, Henry M., *The Twilight of Evolution* (Grand Rapids: Baker Book House,1980)
—*Scientific Creationism* (San Diego: Creation-Life, 1980)
Morris. Henry M. and Whitcomb, John C., *The Genesis Flood* (Philadelphia: Presbyterian and Reformed, 1961)
Sullivan, J.W.N., *The Limitations of Science* (New York: New American Library,1963)
Overman, Dean, *A Case Against Accident and Self-organization* (New York: Rowman & Littlefield, 1997)
Ward, Peter D. and Brownlee, Donald, *The Rare Earth* (New York: Copernicus, 2000)

INDEX

—A—
Abiogenesis, 24
Abraham, vision of worlds, 3; creation account, 95
Adam, subject to Kolob's time,109; "created" on seventh day,116; holds keys of universe, 132
Adam-ondi-Ahman, 157
Adversity, need for, 64
Allen, Daniel, 212, 253, 274, n. 51
America, cities of, inhabited by Lost Tribes, 259
Animals, languages of, 159
Ark, Noah's, 162
Armageddon, battle of, 270
Atomic holocaust, world not to end in, 281
Atonement, the, relation to Fall, 134

—B—
Bacon, Francis, 194
Ballard, Melvin J., 60, 256
Baptism, of Earth,174; spiritual, 279
Benson, Ezra Taft, *viii*
Big Bang, the, 5, 6
Blood, replaced spirit;133; not to be shed wantonly, 178
Body, physical, essential nature of, 56
Brown, Benjamin, 213, 214
Brown, Homer, 213

—C—
Cannon, George Q., 107, 158, 173
Cities, fall of, 270
Clark, J. Reuben, 16, n. 9
Clayton, William, 77, n. 38
Cowdery, Oliver, 209, 251
Cowley, Matthew, 61
Cowley, Matthias, 215
Cain, 143
Canaan, people of, 144
Catastrophism, 28
Celestial, beings, powers of, 311; cities, 312

Chance, life meaningless if product of, 29
Chiasmus, Earth's history plotted as a, 119
Children: alive in Christ, 67; saved in celestial kingdom, 68; have eternal life;70; doctrine on, unclear, 71
Christ: sufferings of, 225; spirit body of, 57; Earth is footstool of, 277; second coming of, 278; exaltation of, 313
Continental shift, 194
Continents, reunion of, 284
Corruption, all, destroyed, 281
Cosmogony, theories of, 6
Council, second creation, 112
Creation: Gods of, 89; place of, 94; dual theory of, 97-98; Abraham and Moses on, 99, 100; time frame of,104; a miracle, 106; summary of, 109-12; spiritual versus physical, 114; chart of, 115
Critchlow, William J., 260

—D—
Darkness, at Christ's crucifixion, 227
Day, use of term by Abraham and Moses, 107; Lord's, 107
Damnation, self, 173
Dead, place of, 82; righteous near, 82; vision of redemption of, 84
Deep, the great, 286
Dibble, Philo, 206
Dove, significance of, 177

—E—
Earth: Moses' visions of, 4; a stewardship, 38; measure of creation, 53; a living, 57; dual organization of, 79; heavenly pattern for, 80; elements of, 91; formation described, 93; organized near Kolob, 94; age of, 102; uniqueness of, 102; dual weeks of (chart), 120; Edenic state of, 131;

mankind's sins against,148; aging of,150; abides celestial law,173; baptism of, 174; division of, 193; lament of, 227 upheavals on, 229; most wicked, 231; temporal span of, 140, 237; to reel, 269; pleads for rest, 277; spiritual baptism of, 279; paradisiacal glory of, 282; dimensions of, 288; future status of, 296; healing of, 289; the twelfth kingdom, 297; death of, 305; resurrection of, 306; return of, 308; a sea of glass, 309; revelatory powers of, 311;celestial cities on, 312; God of, 313; meek to inherit, 314; spirit bodies formed from, 318

Earths, countless, inhabited, 4; parent, 92

Egyptian Alphabet and Grammar,14

Enoch: power of, 145; established City of Holiness, 186; ministry to other planets, 190

Eternity, age of, 104

Eve, 114

Everett, Addison, 211

Evolution, pervasiveness of doctrine, 23; origin of theory, 25; true, 57

—F—

Faith, divine, 40; Lectures on, 41; spoken word of, 42

Fall, the, literal, 134; effects of, 136; subsequent, 142

Famine, 146; effect on Israel, 147

First Presidency, on origin of man, 26; on creation of Earth, 79

Fish, on celestial earth, 310

Flesh, Adam the first, 114

Flood: seen by Enoch, 157; nature of, 163; waters of, 164; reality of, 165; not to be repeated, 167; sequence of (chart), 168; justification for, 172; Earth's baptism, 173

Footstool, Earth is, of Christ, 277

Foreordained, certain spirits, 61

—G—

Galaxy, the, 4

Garden of Eden, created seventh day, 113; nature and location of, 135

Gates, Jacob, 211

Gentiles, not ministered to by Christ, 204

God: dwells in eternal fire, 13; philosophical arguments for, 29; has undergone mortal trials, 64; the greatest chemist, 103; wrath of, 243

Gods, endless, 8

Gondwanaland, 194

Gospel, availability of, 66

Grand council, 61, 90

Grant, Jedediah M., 84

Green, "Sister," 214

Gulf of Mexico, site of Enoch's city, 189

—H—

Heavens, 111

Huntington, Oliver B., 17, n. 19; 121, n. 7; 221, n. 64, 65

Huxley, Julian, 23

—I—

Ice, to melt at presence of Lost Tribes, 254

Intelligences, endless organized, 9

Israel, origin of, 196; remnant of, 200; judgments on, 229; keys of, 251

Ivins, Anthony W., 220, n. 57

—J—

Jesus, birth of, 223; sufferings of, 225; crucifixion of, 226

Johnson, Benjamin F., 206

Judgments: manmade, 241; of God in last days, 243; on Israel in America, 229

—K—

Kepler, Johannes, 11

Kimball, Heber C., 58, 84, 87, 92

Kimball, Spencer W., 256, 261

Knowledge, mankind's limited,19; co-exists with faith, 41

Kolob: seen by Abraham, 13; a governing sphere, 14; location of, 15; site of Grand Council, 90, 108; Earth formed near, 94; measurement of time on, 107

—L—

Land, primal, 185
Language, primeval, 42; in Millennium, 296
Laurasia, 194
Law, natural and spiritual, 20; origin of, 21
Life, sanctity of, 178
Little season, 303
Lost Tribes: origin, 197; led away by Father, 200; journeyings, 201; ministry of Christ to, 203; views on location, 204; a single body, 209; extra-terrestrial location, 210; reconciliation of statements on, 215; Moroni's prophecy on, 250; keys for leading, 251; labors of John, 252; return, 254; prophets among, 255, 261; enemies of, 258; return of, 259; records of, 260

—M—

Mace, Wandle: 188, 212, 216, 257, 264
Man, origin, 27; power, 36; disobedience, 45; procreation,116; fall, 133; longevity, 179; millennial diet, 294
Mankind: rebels against God, 44; is filthiness of earth, 174; millennial nature of, 294
Marriage, purpose of, 58
Matter, receives intelligence, 38; is eternal, 91
McConkie, Bruce R., 19, 208
Meek, inherit Earth, 314
Melchizedek, 192
Methuselah, 145
Mind, nature of divine, 38
Miracles, nature of, 35; of Jesus, 106
Moon, beliefs about, 11
Mortality, pre-mortal knowledge of, 55; a second birth, 56; a proving state, 59

Moses, visions of, 53, 100; creation account, 95
Mountains, ancient,169; present, 194; lowering of, 287

—N—

Nations, bounds of,195; end of all, 270
Naturalism, philosophy of science, *viii*; pure, rejected by scripture, 29
New Jerusalem: union with Enoch,s city, 262; complex of communities, 263; Lost Tribes to come to, 261, 283; built on designated spot, 263; return of, to earth, 312
Nibley, Hugh, 25, 26
Noah: preaching of, 160; next to Adam in authority, 166; is Gabriel, 167; Lord covenanted with,175; given new dietary law, 178
Noble and great, spiritual scholarships of, 61; seen by Abraham, 59
North countries: Lost Tribes come from, 255, 256; great deep driven back to, 283, 284, 286; Ether prophecied of, 312

—O—

Oath: God swore, to Enoch, 277
Odin, identification of, with Christ, 204
Olive leaf, symbolism of, 177

—P—

Pangaea, 194
Papyri, Joseph Smith's, 159
Paradise, 83
Patriarchal order, 53
Patriarchs, life spans of, 179
Partridge, Edward, 211
Peace, millennial, 292
Peleg, 192
Penrose, Charles W., 125, n. 67
Peterson, Mark E., 220, n. 57
Phelps, W.W., 9, 28, 104, 206,
Plan, Abraham's account of, 109
Planets, Earth formed from other, 91; Enoch ministers to other, 190
Plates, tectonic, 171, 194
Power, of spoken word, 46

Pratt, Orson: *ix, x,*11, 20, 21 22, 37, 39, 58, 63, 65, 91, 92, 93, 97, 98, 101, 114, 131, 158, 174, 193, 206, 207, 255, 258, 260, 280, 285, 309, 315

Pratt, Parley P.: 80, 85, 131, 138, 209, 213, 216, 253, 269, 289

Pre-Adamic life, 93

Pre-mortality, foreknowledge in, 55; all tested in, 60

Priesthood: defined, 22; Patriarchal Order of, 53; Earth organized by power of, 106

Prophets, Lost Tribes led by, 255

—R—

Radiation: cosmic,179; possible in last days, 247

Rain: on seventh day, 113

Rainbow: sign of God's oath to Noah,175; sign of Enoch's promised return, 263; departure of, 267

Records, of Lost Tribes, 260

Redeemer, to come to Zion, 265

Resurrection, the last, 304

Rest, Earth pleads for, 227

Restitution, of all things, 283

Resurrection, the last, 304; of Earth, 306

Revelation, source of all knowledge, 36

Reynolds, George: 201, 205

Rigdon, Nancy, 239

Rigdon, Sidney,13

Roberts, B.H, 14, 93, 208

Rogers, Samuel H., 257, 288, 300, n. 39

—S—

Sabbath, Earth's two, 113; lengths of Earth's, 119

Samaritans, 199

Sargon II: deported Israelites to Assyria, 200

Satan, his part in the Fall, 158

Science, limitations of natural, *viii,* 20, 24; powers of modern, 106

Scourge, overflowing, 245; vision of, 247

Scripture: spiritualizing of, vii, *ix*

Sea: covered primal earth, 111; driven back to north countries, 286; boiling of, 281; no, on celestial earth, 310

Seals, the sixth and seventh, 266

Seaton, N.E., 259

Seeds, brought to earth, 112

Senses, source of happiness, 292

Serpent, in Eden, 158

Sickness, desolating, 246

Sign of Son of Man, precedes Christ coming, 264; identified with Enoch's city, 265

Silence, half hour of, 266

Smith, Bathsheba W., 212, 216, 258

Smith, George Albert, 249

Smith, H.M. and Sjodahl, J.M., 207

Smith, Hyrum, 294

Smith, Joseph: *viii, ix, x, xi,* 8, 9, 10, 12,13, 23, 29, 39, 60, 61, 62, 64, 65, 68, 70, 81, 82, 84, 85, 86, 87, 89, 90, 91, 94,103, 104, 105, 108, 117, 135, 140, 146, 149, 159, 162, 187, 190, 192, 195, 205, 206, 209, 211, 212, 214, 216, 225, 232, 238, 239, 245, 248, 250, 251, 253, 254, 255, 257, 258, 259, 261, 263, 264, 265, 267, 269, 270, 281, 282, 283, 284, 287, 292, 296, 297, 304, 308, 309, 310, 311, 313, 316

Smith, Joseph F.: 37, 39, 55, 61, 71, 80, 85, 87, 117

Smith, Joseph Fielding: ll, 21, 60, 67, 98, 114, 209, 216, 232

Smith, William, 104

Snow, Eliza R.: 86, 135, 138, 188, 213, 214, 215, 216, 258, 289

Snow, Lorenzo, 213, 295

Sodom and Gomorrah, 146

Son of Man: sign of, 264; accepted limitations of mortality, 278

Spirit of Lord: governs universe, 22, 39; source of knowledge, 36; prevents chaos, 39; not always with men, 65; withdrawing from earth, 238

Spirits, of just nearby, 82; no infant, 60; transplanted, 85

Spirit world: dead go to, 81; proximity to Earth, 82; a varied society, 83; vision of, 84; dual meaning of, 87
Spiritual death: befell Adam and Eve, 134
Star, new, 224
Stars, eternal,11; to fall, 267
Sun, on being inhabited, 11; fate of, 305

—T—

Talmage, James E., 210, 260
Taylor, John: 22, 94, 172, 191, 213, 297, 314
Time: the unknown factor,102; involved in earth's organization,104; use of, by Abraham,107; pre-Fall, 106; Gods ended labors on seventh,113; earth fell from celestial,136; meridian of, 223; Saturday evening of, 237
Translated beings, 187; extra-terrestrial ministery, 190
Temple, veil of, 228; gathering for purpose of building, 261
Ten tribes: (see Lost Tribes)
Translation, Enoch learned doctrine of, 189
Treasures, of Lost Tribes, 192
Trials, need for, 63-65
Truth, superiority of divine, viii
Twelve, proclamation of the, 193

—U—

Uniformitarianism: 28
Universe, origin of, 6; fate of, 7; the eternal, 8
Urim and Thummim, Earth to be a great, 309

—V—

Vegetation: corrupted by Fall, 140
Vegetarians, all to become, 294
Veil, of temple, 228; Earth must pass through, 298
Vision, of glories, 13; of redemption of dead, 85

—W—

Walker, Charles L., 121, n.10; 211, 214, 215, 222, n. 26; 319, n. 19
War, in heaven, 88; nuclear, 247
Wegener, Alfred, 194
Whitmer, John, 253
Whitney, Orson F., 172, 207, 209
Wickedness, 160
Wisdom, word of, 294
Widtsoe, John A., 19, 24, 58
Woodruff, Wilford, 63, 71, 186, 213, 215, 216, 247, 255, 259
Word, power of spoken, 42
World, moral state of, 240; not to end in atomic holocaust, 281
Worlds, diverse, 10; eternal, 12; numberless, 53

—Y—

Young. Brigham, x, 8, 9,11, 36, 38, 39, 40, 45, 63, 64, 70, 81, 82, 83, 89, 94, 103, 107, 112, 116, 135, 137, 138, 174, 189, 213, 215, 216, 232, 280, 289, 292, 296, 305, 308, 311, 315
Young, Joseph 37, 60, 189

—Z—

Zion of Enoch: translation, 187; location; 189; union with latter-day Zion, 262; identified with sign of Son of Man, 265

Note: With few exceptions, endnotes are not indexed.